CAMBRIDGE LIBRARY COLLECTION

Books of enduring scholarly value

Mathematical Sciences

From its pre-historic roots in simple counting to the algorithms powering modern desktop computers, from the genius of Archimedes to the genius of Einstein, advances in mathematical understanding and numerical techniques have been directly responsible for creating the modern world as we know it. This series will provide a library of the most influential publications and writers on mathematics in its broadest sense. As such, it will show not only the deep roots from which modern science and technology have grown, but also the astonishing breadth of application of mathematical techniques in the humanities and social sciences, and in everyday life.

Odd Numbers

Mathematics has a reputation of being dull and difficult. Here is an antidote. This lively exploration of arithmetic considers its basic processes and manipulations, demonstrating their value and power and justifying an enduring interest in the subject. With humour and insight, the author casts a steady eye on how basic mathematics relates to everyday life – as true now as when this book was originally published in 1940. The introductory treatment of millions, billions and even trillions could be profitably read by aspiring bankers, economists or politicians. H.G. Wells is gently teased for his mistake in applying the law of proportionality in a novel. McKay politely adjusts the astronomical scales selected by the eminent cosmologist Sir James Jeans. He confidently navigates the hazards of averages, approximations and units. For anyone interested in what numbers mean and how they can be used most effectively, this book will still educate and delight.

Cambridge University Press has long been a pioneer in the reissuing of out-of-print titles from its own backlist, producing digital reprints of books that are still sought after by scholars and students but could not be reprinted economically using traditional technology. The Cambridge Library Collection extends this activity to a wider range of books which are still of importance to researchers and professionals, either for the source material they contain, or as landmarks in the history of their academic discipline.

Drawing from the world-renowned collections in the Cambridge University Library, and guided by the advice of experts in each subject area, Cambridge University Press is using state-of-the-art scanning machines in its own Printing House to capture the content of each book selected for inclusion. The files are processed to give a consistently clear, crisp image, and the books finished to the high quality standard for which the Press is recognised around the world. The latest print-on-demand technology ensures that the books will remain available indefinitely, and that orders for single or multiple copies can quickly be supplied.

The Cambridge Library Collection will bring back to life books of enduring scholarly value across a wide range of disciplines in the humanities and social sciences and in science and technology.

Odd Numbers

Herbert McKay

CAMBRIDGE
UNIVERSITY PRESS

CAMBRIDGE UNIVERSITY PRESS

Cambridge New York Melbourne Madrid Cape Town Singapore São Paolo Delhi

Published in the United States of America by Cambridge University Press, New York

www.cambridge.org
Information on this title: www.cambridge.org/9781108002820

© in this compilation Cambridge University Press 2009

This edition first published 1940
This digitally printed version 2009

ISBN 978-1-108-00282-0

ODD NUMBERS

CAMBRIDGE
UNIVERSITY PRESS
LONDON: BENTLEY HOUSE
NEW YORK, TORONTO, BOMBAY
CALCUTTA, MADRAS: MACMILLAN
TOKYO: MARUZEN COMPANY LTD

ODD NUMBERS

OR

Arithmetic Revisited

By Herbert McKay

CAMBRIDGE

AT THE UNIVERSITY PRESS

1940

PRINTED IN GREAT BRITAIN

CINDERELLA IN GLASS SLIPPERS

Arithmetic is usually regarded as the Cinderella of Mathematics, the drudge whose duty it is to do everything that is dull. That is more the fault of arithmeticians than of arithmetic. It is true that a great body of arithmetic has been created that is practically meaningless. We have the dull manipulation of numbers, leading nowhere; the actual results of the manipulation are of no interest to anyone.

But Cinderella has her glass slippers. There are all kinds of interesting things that cry out for arithmetical treatment. A few of them find their way occasionally into the newspapers; but few people seem to have any knowledge of what one may call educated arithmetic, the arithmetic that leads to results that are interesting in themselves.

I have tried to make it clear that the surprising results we are sometimes shown—the apt illustrations given by astronomers and others—are not the result of mathematical magic, but of the ordinary processes of arithmetic. I have used nothing but the straightforward arithmetical processes, the simplest of the trigonometrical ratios, and a little elementary algebra—just those parts of mathematics that are remembered by those who have not specialised in mathematics.

I have to thank several friends who were good enough to help me with the proofs, and especially Mr R. F. Cyster, B.A. and Mr H. A. C. McKay, B.A., who read the whole of them and made a number of helpful and generous criticisms.

H. McK.

January 1940

I

Millions and Billions and Trillions

A MILLION is a very big number. A man with a million pounds is a man to be envied; he is master, or slave, of a very big income. Let us look for one or two comparisons that will help us to realize just how big a million is.

A man's pace is about a yard. A million paces would take him a million yards, or $1,000,000 \div 1760$ miles:

$$
\begin{array}{r}
568 \text{ miles} \\
1760 \overline{)1000000} \text{ yards} \\
8800 \\
\hline
12000 \\
10560 \\
\hline
14400 \\
14080 \\
\hline
320 \\
\hline
\end{array}
$$

So that a million paces would take a man about 568 miles, which is about the distance in a straight line from Land's End to John o' Groats. At 40 miles per day, which is a considerable speed for a long journey, the million paces would take more than 14 days.

Suppose you had a million counters to count out into packets of a hundred. The counters are piled up, more than a foot high, on a table 5 feet by 10 feet. The task is evidently a formidable one. You will probably be able to count one packet in a minute. To find the number of minutes required to count the million we divide by 100, or knock off two noughts:

10,000.

Divide by 60 to change the minutes to hours:

$$60\overline{)10000}$$
$$166 \text{ hr. } 40 \text{ min.}$$

Allowing for an 8 hour day this would be nearly 21 days; and with a 5 day working week the time would be more than 4 weeks.

Pause for a moment to imagine yourself working hour after hour at this perpetual counting, day after day, and week after week, until the last hundred counters had been put into its packet. You may begin to have a considerable respect for big numbers, and to look at the string of six o's in 1,000,000 with some sense of what they indicate.

A million days! When we wish a man long life we sometimes cry out "May he live a thousand years!" An extravagant wish, but it would be still more extravagant to wish him a million days.

$$365\overline{)1000000} \text{ days} \quad 2740 \text{ years}$$
$$\underline{730}$$
$$2700$$
$$\underline{2555}$$
$$1450$$
$$\underline{1460}$$

A million days is just about 2700 years. (For this purpose we need not trouble about leap years, or getting the end figures exactly.) It is not quite three-quarters of a million days since Julius Caesar landed in Britain and found it peopled with blue-painted Celts.

Here is a tiny cube with its edges one-tenth of an inch long. Suppose a million of them were placed side by side. They would stretch 100,000 inches or

$$100,000 \div 12 = 8333\tfrac{1}{3} \text{ feet,}$$

which is well over a mile and a half. Imagine a mile and a half of these tiny cubes!

We can arrange the cubes to take up less space. Look at this square. It contains ten rows of small cubes each holding ten, that is $10 \times 10 = 100$ alto- 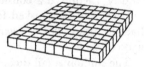 gether. To make up a million we should want 10,000 of these one-inch squares. $10,000 = 100 \times 100$. So we should want a space 100 inches long and 100 inches wide, completely covered with the tiny dice. The total number of dice is

$$1000 \times 1000 = 1,000,000.$$

(There are 1000 rows each of 1000 dice.)

Draw on a wall a square with sides 8 ft. 4 in. long. Divide the sides into inches and join across; there are 198 of these lines to draw across the square, besides the 4 outside lines—560 yards of lines. We now have the big square divided into 1 inch squares. Begin at the top left-hand corner, divide the first square into a hundred small squares each with sides a tenth of an inch long. Then proceed to the next. And so on.

What a task! It would be futile to ask anyone to perform it. It might be sufficient to divide one of the 1 inch squares into hundredths, and then to realize that there are still 9999 to be done.

A film firm with which I was connected was asked to represent a 40,000 horse-power turbine by means of a drawing showing 40,000 horses simultaneously on the screen. When we politely pointed out the absurdity of this, the turbine firm insisted on 40,000 dots. The artist was accordingly told to draw 40,000 dots. He drew 400 dots, and then refused to draw any more when I pointed out to him that he had only accomplished one-hundredth of his task. Finally the drawing of 400 dots was reproduced as a block, a hundred careful prints were made from it, and these were pasted in a square. Even this was a long and tedious job. The word "dotty" acquired a considerable vogue in those days.

If we had been asked for a million dots we should have had to paste down twenty-five times as many squares, 2500 of them.

The million small dice, that stretched a distance of more than a mile and a half when placed side by side, were tucked into a square that was only 8 ft. 4 in. wide. Let us see how we can tuck them away still more snugly. Look at this 1 inch cube. Look at the front layer of small dice; there are 10 rows of 10; and altogether there are 10 layers. So we have $10 \times 10 \times 10 = 1000$ small dice in the cube. We have packed 1000 dice into the small space of an inch cube.

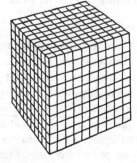

To make up a million dice we should want 1000 of these inch cubes, that is a thousand thousand dice, or $1000 \times 1000 = 1,000,000$. We can pack the inch cubes in the same way, into a bigger cube 10 inches by 10 inches by 10 inches. In each edge of this bigger cube there are 100 dice, so that there are 100 × 100 = 10,000 dice in each of 100 layers. That is 100 × 100 × 100 = 1,000,000 altogether.

See again what we have done. Placed side by side a million tenth-of-an-inch dice stretch more than a mile and a half. Arranged in a square they fill a space 8 ft. 4 in. square. Arranged in a cube they occupy a cube with 10 inch edges. It really is extraordinary how the biggest numbers climb down when we begin to arrange them. It is all because a million is 1,000,000 × 1, or 1000 × 1000, or 100 × 100 × 100.

The distance round the world is 25,000 miles—quite a considerable distance in its way. An express train running at 50 miles per hour would take 25,000 ÷ 50 = 500 hours, or nearly 21 days to do this distance. But the distance is a mere fortieth part of a million miles. Steaming ceaselessly night and day at 50 miles per hour an express train would take 20,000 hours, or 833 days, or nearly 2⅓ years to go a million miles.

The moon is nearly ten times as far away as the distance round the earth—238,000 miles (not far short of a quarter of a million miles); so that our express train would take more than six months to perform a journey equal to the distance of the moon.

The distance of the sun is about 93 million miles. Let us see how many times greater that is than the distance of the moon: 93,000,000 ÷ 238,000 = 93,000 ÷ 238

$$\begin{array}{r} 390 \\ \hline 238)\overline{93000} \\ 714 \\ \hline 2160 \\ 2142 \\ \hline \end{array}$$

We need not trouble about the end figure. The sun is about 390 times as far off as the moon. To travel the distance to the sun our express train would take:

$$93,000,000 \div 50 = 1,860,000 \text{ hours,}$$
$$= 77,500 \text{ days,}$$
$$= \text{about } 212 \text{ years.}$$

So we have the comparison:

Round the earth: 25,000 miles— 21 days.
To the moon: 238,000 miles— 6 months.
To the sun: 93,000,000 miles—212 years.

Here is another comparison for these distances. Sound travels with extreme rapidity when compared with a railway train. The speed is about a mile in 5 seconds, as every child knows who has counted the distance of a lightning flash. A mile in 5 seconds is 12 miles per minute or 720 miles per hour. To go the distance round the world sound would take $25,000 \div 720 =$ about 35 hours, so that sound would take nearly a day and a half to travel the distance round the world.

No sound can ever reach us from the sun or the moon, because there is no air to carry the sound. But suppose there were, how long would sound take to reach us from the moon? The answer is easy enough:

$$238,000 \div 720 \text{ hours}$$

$$\begin{array}{r} 330 \text{ hours} \\ 72\overline{)23800} \\ 216 \\ \hline 220 \\ 216 \\ \hline 40 \end{array}$$

(The end figure does not matter.) 330 hours is 13¾ days.
So that if we *could* hear moon sounds we should hear them
13¾ days after they were made.

And what about the sun?

$$
\begin{array}{r}
129000 \text{ hours} \\
72\overline{)\,9300000} \\
72 \\
\hline
210 \\
144 \\
\hline
660 \\
648 \\
\hline
\end{array}
$$

(There is no need to carry the division further; indeed we
might have called the number 130,000.)

129,000 hours = about 5000 days = about 14 years.

If we *could* hear explosions in the sun we should hear them
14 years after they happened.

Millions do very well for the larger affairs of everyday life.
The population of Greater London is more than 8 millions,
Greater New York is about the same, and we begin to consider a town really large when it passes the million mark.
Just as we consider a man really wealthy when he becomes
a millionaire. National revenues are reckoned in hundreds
of millions. Distances in the solar system are equally well
expressed in millions of miles, from the quarter of a million
miles distance of the moon from the earth, to the 2794
millions of miles that Neptune lies out from the sun. A
million, with its string of six noughts, is indeed an admirable
and useful number.

And now for a jump! We jump suddenly from a million

to a billion, from 1,000,000 to 1,000,000,000,000. The English (and German) billion is a million times a million, a million squared as we say; the number of noughts is doubled, a billion has twelve noughts. In the United States and on most of the Continent a billion is a mere thousand millions —1,000,000,000, the thousandth part of our billion. Allowing five dollars to the pound a dollar billionaire has the equivalent of 200 million pounds. A sterling billionaire would have to possess the impossible sum of a billion pounds; if he could exist he could buy out five thousand dollar billionaires. Which shows that there is something in a name.

Millions do well enough for distances in the solar system. When we come to star distances millions are no longer big enough; we need billions. The distance of the nearest star is 26 billion miles. No human mind can realize what a billion miles means, but we can get some faint idea of the immensity of the distance by finding how long light takes to come from the star.

Light is the speediest thing we know of. Its speed is so great that in ordinary conditions it seems to take no time at all to move from one place to another; it is only across immense distances that the time it takes becomes apparent. The speed of light has been measured many times and in many ways. The results agree in giving the speed as about 186,000 miles per second.

The immensity of that speed may be judged by comparing it with the speed of an express train. At 50 miles per hour an express train would do 186,000 miles in

$$186,000 \div 50 \text{ hours,}$$
$$= 3720 \text{ hours,}$$
$$= 155 \text{ days,}$$
$$= 5 \text{ months of } 31 \text{ days.}$$

So that a train, travelling night and day continuously at 50 miles per hour, would take five months to travel the distance that light leaps across in a second.

$$186,000 \div 25,000 = \text{about } 7\tfrac{1}{2}.$$

So that light travels in a second a distance equal to $7\tfrac{1}{2}$ times the distance round the world. To come from the moon, light takes $238,000 \div 186,000 = \text{about } 1\tfrac{1}{4}$ seconds. To come from the sun it takes $93,000,000 \div 186,000 = 500$ seconds, or rather more than 8 minutes.

How far does light travel in a year? It sounds as though the number of miles would be extravagantly large, but the calculation is easy enough.

186,000 miles in a second,
$= 186,000 \times 3600$ miles in an hour,
$= 186,000 \times 3600 \times 24$ miles in a day,
$= 186,000 \times 3600 \times 24 \times 365\tfrac{1}{4}$ miles in a year.

When those four numbers are multiplied out we get a number that is about 5·9 billion miles. This distance is called a light-year. A light-year is a length, not a time; it is the distance that light travels in a year; it is about 5·9 billion miles.

When we give the distance of a star we often express it in light-years. The distance of the nearest star (α Centauri—the brightest star in the Centaur in the southern heavens) is 4·4 light years. The North Star is 47 light-years away. And there are other stars hundreds and thousands of light-years from us. If for any reason we want to turn light-years into billions of miles we have only to multiply the number of light-years by 5·9. If we know the distance in billions of miles we can change it into light-years by dividing by 5·9.

A billion miles is an admirable unit for star distances. We use the light-year because of the imaginative suggestion of light, leaping at almost incredible speed across immense distances, taking four and a half years to reach us at 186,000 miles a second.

We saw how a million tenth-inch dice could be arranged to form a line more than a mile and a half long, or compressed into a square with 8 ft. 4 in. sides, or tucked away snugly into a cube with 10 inch edges. We can do the same sort of things with a billion tenth-inch dice. Spread out in a line we should have a line a million times as long—the line would stretch about one and a half million miles, or sixty times round the world. The sides of the square would be a thousand times as long as 8⅓ feet—more than a mile and a half long; think of an immense square field—about 1600 acres—and then think of the whole of this immense space divided into small squares with sides a tenth of an inch long. The cube would have edges a hundred times as long as the 10 inch cube used for a million dice:

$$10 \times 100 \text{ in.} = 1000 \text{ in.}$$
$$= 83\tfrac{1}{3} \text{ feet.}$$

We have seen how numbers climb down when we arrange the objects in squares, and still more when we arrange them in cubes. Now let us see how they jump up when we multiply.

A mile is 5280 feet—a quite comprehensible number.

A square mile is 5280^2 or 5280×5280 square feet $= 27,878,400$ square feet, or nearly 28 million square feet.

A cubic mile is 5280^3 or $5280 \times 5280 \times 5280$ cubic feet $= 147,197,952,000$ cubic feet or more than a seventh of a billion cubic feet. A big enough number! But then you must remember what a cubic mile is. We have to think of an immense field with sides a mile long, and we have to think of this space covered with 1 foot cubes. Then we have to think of 5280 of these layers piled one on another.

If we know the radius of a sphere we can easily find its volume. The volume is:

$$\tfrac{4}{3} \times 3\tfrac{1}{7} \times \text{radius}^3.$$

We find the cube of the radius, and then multiply this by $\tfrac{4}{3}$ and $3\tfrac{1}{7}$. The radius of the earth is 4000 miles, and so its volume is:

$\tfrac{4}{3} \times 3\tfrac{1}{7} \times 4000^3$ cubic miles
    = about 268,000,000,000 cubic miles,

or more than a quarter of a billion cubic miles.

What is the mass of the earth in tons?

We know that the earth is $5\tfrac{1}{2}$ times as heavy as water, and that a cubic foot of water weighs about $62\tfrac{1}{2}$ pounds. Well,

268,000,000,000 cubic miles
    = 268,000,000,000 × 147,000,000,000 cubic feet.

If this were water we could find its weight in pounds by multiplying by $62\tfrac{1}{2}$. So that the earth weighs in pounds:

268,000,000,000 × 147,000,000,000 × $62\tfrac{1}{2}$ × $5\tfrac{1}{2}$,

or in tons the same number divided by 2240
    = about 6,000,000,000,000,000,000,000 tons.

A million has 6 noughts.
A billion has 12 noughts.
A trillion has 18 noughts.
So that the mass of the earth is about 6000 trillions of tons.

It would be futile to continue the method we have used for arranging a million and a billion small dice. Placed side by side the dice would stretch something like one and a half billion miles, which is no help at all. The sides of a square to contain a trillion of the small dice would be more than 1500 miles long. Even a cube to contain a trillion of the small dice would have edges more than a mile and a half long.

It is difficult if not impossible to imagine a cube with edges a mile and a half long; it is the height that is the trouble. Suppose we spread out a trillion of these small dice so that they cover the ground to a depth of a foot. The edge of a foot cube contains 120 small dice, so that a cubic foot contains 120^3 dice. (We need not multiply out 120^3 just yet.) A trillion of the dice would occupy:

$$\text{a trillion} \div 120^3 \text{ cubic feet.}$$

As they are being spread out a foot deep they will cover an area of the same number of square feet. To bring the square feet to square miles we divide by 5280^2.

And so we find that a trillion of the small dice would cover to a depth of 1 foot an area of:

$$\frac{1,000,000,000,000,000,000}{120^3 \times 5280^2} \text{ square miles.}$$

Let us work this out.

$$120^3 = 120 \times 120 \times 120 = 12 \times 12 \times 12 \times 10 \times 10 \times 10$$
$$= 1,728,000.$$
$$5280^2 = 5280 \times 5280 \quad = 27,878,400.$$
$$120^3 \times 5280^2 = 1,728,000 \times 27,878,400$$
$$= \text{(about)} \ 1,700,000 \times 28,000,000$$
$$= 47,600,000,000,000$$
$$= \text{(about)} \ 48,000,000,000,000.$$
$$\frac{1,000,000,000,000,000,000}{48,000,000,000,000}$$
$$= \frac{1,000,000}{48}$$
$$= 20,000 \text{ (near enough).}$$

So that a trillion of the small dice would cover to a depth of one foot an area of about 20,000 square miles.

We can find a suitable comparison by looking in a reference book for a country with an area of about 20,000 square miles. The area of Scotland is about 30,000 square miles. So that a trillion of these tenth-inch dice would cover two-thirds of Scotland to a depth of a foot or the whole of Scotland to a depth of eight inches.

The area of England is about 50,000 square miles ($2\frac{1}{2}$ times what we want), and so a trillion of the dice would cover England to a depth of $12 \div 2\frac{1}{2} =$ about 5 inches.

It might seem that the area of the whole world in square miles would work out at an enormous number. If we know the radius of a sphere we can readily find its surface area. This is:

$$4 \times 3\frac{1}{7} \times radius^2$$

$$= 12\frac{4}{7} \times radius^2$$

$$= (about) \ 12\frac{1}{2} \times radius^2.$$

$$4000^2 = 16,000,000 \ (square \ miles).$$

Multiplying this by $12\frac{1}{2}$ we get:

$$200,000,000 \ square \ miles.$$

The estimated area of the world is actually a little less than this (the radius is rather less than 4000 miles). It is about $196\frac{1}{2}$ million square miles. Of this area $55\frac{1}{2}$ million square miles is land and 141 million square miles water.

The population of the world is in the region of 2000 millions—perhaps rather more than this. So that the average density of population is $2000 \div 55\frac{1}{2} =$ about 36 to the square mile.

Before leaving these big numbers it may be noted that the choice of meanings to be attached to billion and trillion is an example of the arithmetical genius of the English people. Million, billion, and trillion are all three convenient units. Millions of pounds express in a most convenient form great wealth and large amounts of capital. We have the higher unit of a hundred million pounds for stating national revenues, and the still higher unit of a thousand millions for stating national debts and national capital values. The populations of countries and of the largest cities are conveniently expressed in millions. Distances in the solar system are conveniently expressed in millions of miles: we remember the distance of the moon as $\frac{1}{4}$ million miles, of the sun as 93 million miles, and so on. Large areas on the earth are conveniently given in millions of square miles. (The reason is that the radius is 4 thousand miles, and the square of a thousand is a million.) The area of the Pacific is about 64 million square miles, of the Atlantic $31\frac{1}{2}$ million, of Asia 17 million, and so on.

For star distances a billion miles is a convenient unit, without the use of the light-year; and a trillion for such measurements as the mass of the earth in tons.

II

Great Powers and Little Powers

THE idea of *continuity* is an essential part of arithmetic. The simplest example of it is the series of numbers: 1, 2, 3, 4, 5, 6, and so on. We go on adding one each time. There are no disconcerting breaks. We do not jump, for example, from 98 to 100 (with no such number as 99).

If we go backward we must have the same sort of continuity, taking off one at a time:

$$6, 5, 4, 3, 2, 1, ?$$

What is the number below 1? 1 less than 1 is 0, so we write 0 as the next number. What is the number below 0? 1 less than 0 (1 in debt, if you like) is written − 1. And so we get the series:

$$6, 5, 4, 3, 2, 1, 0, -1, -2, -3,$$

which may be continued indefinitely at each end.

In reckoning years there is an awkward discontinuity between A.D. 1 and B.C. 1; there was no year 0. The years should run:

A.D. 3	A.D. 2	A.D. 1	0	B.C. 1	B.C. 2	B.C. 3
3	2	1	0	−1	−2	−3

but the year 0 was omitted. From A.D. 1 back to B.C. 1 is not two years, but only one. From A.D. 20 back to B.C. 10 is not 30 years, but only 29. This discontinuity caused the Italians to celebrate the bimillenary of Augustus, founder of the Roman Empire, a year too soon. However, they made up for the error by continuing the celebrations for more than a year.

In arithmetic we carefully avoid such discontinuities; otherwise we could never be sure of our results. We shall consider now an example which shows how continuity is preserved.

Anyone might be excused for getting tired of writing down long numbers—of writing the string of 18 noughts in a trillion, for example. There is a shorter way of writing such numbers. We write 100 as 10^2 which means 10×10. The little 2 is called an index; it tells how many 10's are multiplied together. 4^2 means 4×4, 8^2 means 8×8, and so on. It is easy to continue the process upwards. 10^3 means $10 \times 10 \times 10$ (3 tens multiplied together). 10^4 means $10 \times 10 \times 10 \times 10$; and so on. A million is $10^6 = 10 \times 10 \times 10 \times 10 \times 10 \times 10$. A billion is 10^{12} and a trillion is 10^{18}.

Now let us go backwards:

$$10^3,\ 10^2,\ 10^1,\ 10^0,\ 10^{-1},\ 10^{-2},\ 10^{-3},$$

and so on. But what does 10^1 mean? What on earth does 10^0 mean? (No tens multiplied together!) And what do 10^{-1}, 10^{-2}, 10^{-3} mean?

We have to find meanings for these expressions that preserve arithmetical continuity.

10^4 (the fourth power of 10) $= 10{,}000$

$$10^3 = 1000$$
$$10^2 = 100$$
$$10^1 = ?$$
$$10^0 = ?$$
$$10^{-1} = ?$$

On the left the indices decrease 1 by 1 : 4, 3, 2, 1, 0, −1. On the right each number after the first is a tenth of the one above it—1000 is a tenth of 10,000; 100 is a tenth of 1000. The next number on the right must be a tenth of 100, to preserve the continuity, that is 10. So we say that $10^1 = 10$. The next number on the right must be a tenth of 10, that is ·1. So we say that $10^0 = 1$; we have found a meaning for the apparently nonsensical 10^0 that fits in with our idea of continuity. The next number on the right must be a tenth of 1, that is $\frac{1}{10}$. So we say that $10^{-1} = \frac{1}{10}$ or $\frac{1}{10^1}$. Going on in the same way we find:

$$10^{-2} = \frac{1}{100} = \frac{1}{10^2}$$

$$10^{-3} = \frac{1}{1000} = \frac{1}{10^3}$$

and so on. If we remember that we are keeping the continuity we can say confidently that:

$$10^{-37} = \frac{1}{10^{37}}$$

We have been dealing with powers of 10. Exactly the same idea of continuity is used with the powers of any number:

$$4^3 = 4 \times 4 \times 4 = 64$$
$$4^2 = 4 \times 4 \quad\ = 16$$
$$4^1 \qquad\qquad = 4$$
$$4^0 \qquad\qquad = 1$$
$$4^{-1} \qquad\quad = \tfrac{1}{4}$$
$$4^{-2} = \frac{1}{4^2} \qquad = \frac{1}{16}$$

and so on.

MCK

Any number to the power $0 = 1$. We can express this by writing $x^0 = 1$, where x stands for any number. Also $x^{-1} = \dfrac{1}{x^1}$, $x^{-2} = \dfrac{1}{x^2}$, and so on.

Now let us look at another example of continuity:

$$10^2 \times 10^3 = (10 \times 10) \times (10 \times 10 \times 10)$$
$$= 10^5 \ (5 \text{ tens multiplied together})$$
$$= 10^{2+3}$$

It is easy to see that we add the indices when we multiply.

$$10^5 \times 10^4 = 10^{5+4}$$
$$= 10^9, \quad \text{and so on.}$$

$x^a \times x^b = x^{a+b}$, no matter what numbers x, a and b stand for.

$$10^2 \times 10^9 = 10^{11}$$
$$5^3 \times 5^4 = 5^7, \quad \text{and so on.}$$

We can equally well say:

$$10^2 \times 10^3 \times 10^5 = 10^{2+3+5} = 10^{10}$$
$$10^6 \times 10^{-2} \qquad = 10^{6-2} \ = 10^4$$
$$10^3 \times 10^{-7} \qquad = 10^{3-7} \ = 10^{-4}$$
$$10^2 \times 10^{-2} \qquad = 10^{2-2} \ = 10^0$$
$$10^{-2} \times 10^{-4} \quad = 10^{-2-4} = 10^{-6}$$

Let us check one or two of these; we want to be perfectly sure that there are no discrepancies.

$$10^{-2} = \frac{1}{10^2}, \text{ so } 10^6 \times 10^{-2} = \frac{10^6}{10^2} = 10^4$$

$$10^{-7} = \frac{1}{10^7}, \text{ so } 10^3 \times 10^{-7} = \frac{10^3}{10^7} = \frac{1}{10^4} = 10^{-4}$$

$$10^2 \times 10^{-2} = \frac{10^2}{10^2} = 1 = 10^0$$

Let us set the last line out differently:

$$10^2 \times 10^{-2} = 10^2 \times \frac{1}{10^2} = \frac{10 \times 10}{10 \times 10} = 1 = 10^0$$

What about division?

$$10^5 \div 10^2 = \frac{10 \times 10 \times 10 \times 10 \times 10}{10 \times 10}$$

$$= 10 \times 10 \times 10$$

$$= 10^3$$

$$= 10^{5-2}$$

It is easy to see that we cancel two of the tens, and that we can get the result straight away by subtracting the index of the number we divide by. We should therefore be able to get the result in every case by subtracting the index of the number we divide by:

$$10^6 \div 10^2 = 10^{6-2} = 10^4$$

$$10^3 \div 10^1 = 10^{3-1} = 10^2$$

$$10^2 \div 10^8 = 10^{2-8} = 10^{-6}$$

$$10^{-3} \div 10^2 = 10^{-3-2} = 10^{-5}$$

$$10^4 \div 10^4 = 10^{4-4} = 10^0 \quad (= 1\text{—of course!})$$

There is a little difficulty when we come to an example like:

$$10^{-3} \div 10^{-4} = 10^{-3-(-4)} = ?$$

What does $-(-4)$ mean? On the principle of continuity minus must always mean the same thing. It means the opposite of $+$. -4 is the opposite of $+4$. If $+4$ means 4 yards up, then -4 means 4 yards down. If $+4$ means 4 feet to the right then -4 means four feet to the left. If $+4$ means £4 in credit, then -4 means £4 in debt. And so on. $-(-4)$ means the opposite of -4, that is to say it means $+4$.

2-2

Then: $$10^{-3} \div 10^{-4} = 10^{-3-(-4)}$$
$$= 10^{-3+4}$$
$$= 10$$

We can check this result as follows:

$$10^{-3} \div 10^{-4} = \frac{1}{10^3} \div \frac{1}{10^4}$$

$$= \frac{1}{10^3} \times \frac{10^4}{1}$$

$$= 10$$

We have been talking almost entirely about powers of 10, but the ideas apply equally well—of course they do—to powers of other numbers.

$$6^{-3} \times 6^{-4} = 6^{-3-4} = 6^{-7}$$

$$14^6 \div 14^{-2} = 14^{6-(-2)} = 14^8$$

$$(27\tfrac{1}{4})^{-3} \div (27\tfrac{1}{4})^{-4} = (27\tfrac{1}{4})^{-3-(-4)} = (27\tfrac{1}{4})^{-3+4} = 27\tfrac{1}{4}$$

We have to be careful not to do anything so absurd as to write:
$$2^3 \times 5^2 = 2^{3+2} = 2^5$$

Because $$2^3 \times 5^2 = 2 \times 2 \times 2 \times 5 \times 5 = 200,$$

and not $$2 \times 2 \times 2 \times 2 \times 2 = 32.$$

We can use the device of adding or subtracting indices only when we are multiplying or dividing powers of the same number.

What does $(10^3)^2$ mean? It is the square of 10^3.

$$(10^3)^2 = 10^3 \times 10^3$$
$$= 10^6$$
$$(10^4)^3 = 10^4 \times 10^4 \times 10^4$$
$$= 10^{12}$$

It is easy to see that in finding powers of powers we have to multiply the indices. We have to do this in every case:

$$(10^5)^6 \quad = 10^{5 \times 6} \quad = 10^{30}$$

$$(10^2)^{-3} \quad = 10^{2 \times (-3)} = 10^{-6}$$

$$(10^{-3})^{-4} = 10^{-3 \times (-4)} = 10^{12}$$

We can check the last example as follows:

$$(10^{-3})^{-4} = \frac{1}{10^{-3} \times 10^{-3} \times 10^{-3} \times 10^{-3}}$$

$$= \frac{1}{10^{-12}}$$

$$= 10^{12}$$

What does 10^{2^2} mean? It means 10 to the power 2^2, that is $10^4 = 10,000$.

$10^{10^{10}}$ means 10 to the power 10^{10} or 10 to the power 10,000,000,000, that is:

$$10^{10,000,000,000}$$

That is to say 1 followed by 10,000,000,000 noughts. If that number were written in full, allowing a tenth of an inch to each figure it would stretch 1,000,000,000 inches, = about 16,000 miles, nearly two-thirds of the distance round the world.

Really $10^{10^{10}}$ stands for an absurdly big number. $10^{10^{10^{10}}}$ hardly bears thinking about. The mere number of noughts in writing it down would be equal to that frightful number that stretches 16,000 miles. We had better stop before we get to $10^{10^{10^{10^{10}}}}$. It does show, however, where you may land yourself if you begin to multiply recklessly.

III

How we got Logarithms

ALL the indices in the last chapter were whole numbers,
some of them positive and some negative. Is it possible to
have fractional indices? We can of course write down $10^{\frac{1}{2}}$
(10 to the power $\frac{1}{2}$); but has the expression any sensible
meaning? The meaning, if any, must fit in with the universal
rule of adding indices to multiply powers of a number, and
subtracting when we divide. It would be futile to find a
meaning that did not fit in—the admission of such a meaning
would blow the principle of continuity sky-high.

Following the universal rule:

$$2^{\frac{1}{2}} \times 2^{\frac{1}{2}} \text{ should equal } 2^{\frac{1}{2}+\frac{1}{2}} = 2^1 = 2.$$

Now the square root of 2 multiplied by itself is equal to 2
(that is what the square root means).

$$\sqrt{2} \times \sqrt{2} = 2,$$
$$2^{\frac{1}{2}} \times 2^{\frac{1}{2}} = 2.$$

So that $2^{\frac{1}{2}} = \sqrt{2}$. We have found a meaning for $2^{\frac{1}{2}}$ that fits in
with the universal rule.

$$2^{\frac{1}{3}} \times 2^{\frac{1}{3}} \times 2^{\frac{1}{3}} = 2^{\frac{1}{3}+\frac{1}{3}+\frac{1}{3}} = 2,$$

and

$$\sqrt[3]{2} \times \sqrt[3]{2} \times \sqrt[3]{2} = 2,$$

so that

$$2^{\frac{1}{3}} = \sqrt[3]{2}.$$

We can continue with: $2^{\frac{1}{4}} = \sqrt[4]{2},$

$$2^{\frac{1}{5}} = \sqrt[5]{2}, \quad \text{and so on.}$$

And of course we might have any other number instead of 2.

$$10^{\frac{1}{2}} = \sqrt{10},$$

$$9^{\frac{1}{3}} = \sqrt[3]{9}, \quad \text{and so on.}$$

We have found meanings for $2^{\frac{1}{2}}$, $2^{\frac{1}{3}}$, $2^{\frac{1}{4}}$, and so on, that fit in with the universal rule of indices. We can therefore use these fractional indices in the same sort of way that we use whole number indices:

$$10^{\frac{1}{2}} \times 10^{\frac{1}{3}} = 10^{\frac{1}{2}+\frac{1}{3}} = 10^{\frac{5}{6}}$$

And now we have to find a meaning for $10^{\frac{5}{6}}$, a meaning that fits in with the universal rule. We use nothing but what we have already found to fit in with the rule:

$$10^{\frac{5}{6}} = 10^{\frac{1}{6}+\frac{1}{6}+\frac{1}{6}+\frac{1}{6}+\frac{1}{6}}$$
$$= 10^{\frac{1}{6}} \times 10^{\frac{1}{6}} \times 10^{\frac{1}{6}} \times 10^{\frac{1}{6}} \times 10^{\frac{1}{6}}$$
$$= (10^{\frac{1}{6}})^5$$
$$= (\sqrt[6]{10})^5$$

So that $10^{\frac{5}{6}}$ stands for the fifth power of the sixth root of 10. Somehow or other we should have to find the sixth root of 10, and then raise this number to the fifth power (that is multiply together five of these numbers). Here is an easy example:

$$64^{\frac{5}{6}} = \sqrt{64} \times \sqrt[3]{64} = 8 \times 4 = 32,$$

and $$64^{\frac{5}{6}} = (\sqrt[6]{64})^5 \quad = 2^5 \quad = 32.$$

It is possible to write any number as a power of 10, not quite exactly in most cases, but as exactly as we want. That is to say we can find a power of 10 to equal any number.

Which power of 10 is equal to 2? We know it is a fractional power, because 2 is less than 10.

$2^1 = 2$; $2^2 = 4$; $2^3 = 8$; $2^4 = 16$; $2^5 = 32$; $2^6 = 64$; $2^7 = 128$; $2^8 = 256$; $2^9 = 512$; $2^{10} = 1024$. Now 1024 is not much greater than 1000 ($= 10^3$).

$$2^{10} \text{ is a little greater than } 10^3$$

Let us take the 10th root of each number:

$$(2^{10})^{\frac{1}{10}} = \text{a little more than } (10^3)^{\frac{1}{10}}$$

$$2^{\frac{10}{10}} = \text{a little more than } 10^{\frac{3}{10}}$$

That is $2 = \text{a little more than } 10^{.3}$

$$= 10^{.3+}$$

(When $+$ is added like this it means "a little more than".)

This is not an accurate result, because 1024 is more than 2 per cent more than 1000. We can get a more accurate result by continually doubling until we find a number that is more accurately equal to a power of 10. If we went to the enormous labour of raising 2 to the power 93 we should get a number rather less than 10^{28} (1 followed by 28 noughts). Hence:

$$2^{93} = \text{(a little less than) } 10^{28}$$

Taking the 93rd root of each side:

$$2 = \text{(a little less than) } 10^{\frac{28}{93}}$$

$$= \text{(a little more than) } 10^{.3010}$$

That is a fairly accurate result, but the amount of labour required to find it is enormous. It does, however, show that it is possible to get fairly accurate results even by such crude methods. We shall see later that there are much easier methods of getting results that are just as accurate as we want.

Even with the simpler methods of calculating the powers of 10 there is no need to go to the labour of calculating each power. Suppose we happen to get good results for 2 and 3:

$$2 = 10^{.30103}$$
$$3 = 10^{.47712}$$

Then
$$4 = 2^2 = 10^{.30103 \times 2} = 10^{.60206}$$
$$5 = 10 \div 2 = 10^{1 - .30103} = 10^{.69897}$$
$$6 = 2 \times 3 = 10^{.30103 + .47712} = 10^{.77815}$$
$$8 = 2^3 = 10^{.90309}$$
$$9 = 3^2 = 10^{.95424}$$

We can use the two results (for 2 and 3) also for $12 = 2^2 \times 3$, $15 = 3 \times 10 \div 2$ (or 3×5, since we have found the power for 5), $16 = 2^4$, $18 = 2 \times 3^2$, $20 = 2 \times 10$, and so on. It may be seen that it is only necessary to calculate for the prime numbers.

A very valuable point about powers of 10 is that we can readily find the whole number in the index. As soon as we know that $2 = 10^{.30103}$ we can change to powers of 10—20, 200, 2000, ·2, ·02, etc., that is to say 2 multiplied by any power of 10.

$$200 = 2 \times 10^2$$
$$= 10^{.30103} \times 10^2$$
$$= 10^{2.30103}$$
$$·2 = 2 \times 10^{-1}$$
$$= 2^{.30103} \times 10^{-1}$$
$$= 2^{-1 + .30103}$$

(We often write this as $2^{\bar{1}.30103}$. The minus over 1 shows that the 1 only is negative; the decimal part is positive.)

If we know that $2·367 = 10^{.3742}$ we can say at once that

$236 \cdot 7 = 10^{2 \cdot 3742}$. It is not difficult to see that the whole number part of the index is 1 less than the number of figures before the decimal point. In the tables it is only necessary to give the decimal part of each index.

To show the power of this method of changing numbers to their equivalent powers of 10, we will find the value of 2^{64}. We could multiply together 64 twos, but the process would be terribly long and tedious. But:

$$2^{64} = \left(10^{\cdot 30103}\right)^{64}$$
$$= 10^{\cdot 30103 \times 64}$$
$$= 10^{19 \cdot 2659}$$

The 19 tells us that there are 20 figures before the decimal point. A book of tables tells us that $10^{\cdot 2659} = 1 \cdot 845$.

Hence $2^{64} = 18,450,000,000,000,000,000$.

When indices are used in this way they are called logarithms. When we say that the logarithm of 2 is $\cdot 30103$, or that:

$$\log 2 = \cdot 30103,$$

this is another way of saying that:

$$2 = 10^{\cdot 30103}$$

10 is called the base of the logarithms; we sometimes put it in like this:

$\log_{10} 2 = \cdot 30103$ (the logarithm of 2 to base $10 = \cdot 30103$).

It is not necessary to write logarithms as indices. We can say:

$$\log 2^{64} = 64 \log 2$$
$$= 64 \times \cdot 30103$$
$$= 19 \cdot 2659$$

And $19 \cdot 2659$ is the logarithm of $18,450,000,000,000,000,000$.

There is an old problem which can readily be solved by using logarithms.

If Adam had invested a penny (the lowest amount a bank recognizes) at a half of one per cent compound interest, what would it amount to after 6000 years?

At $\frac{1}{2}$ per cent interest £100 becomes £100·5 in a year. So that each pound becomes £1·005 (we divide by 100), and this happens yearly. Adam's penny (which is £$\frac{1}{240}$) amounts in 6000 years to:

$$£\tfrac{1}{240} \times 1·005^{6000}$$

We will work this out first by using the logarithms as indices, and then in the usual way.

(i) $\dfrac{1}{240} \times 1·005^{6000} = \dfrac{1}{10^{2·38021}} \times 10^{·00217\times6000}$

$$= 10^{·00217\times6000-2·38021}$$

$$= 10^{13·02-2·38}$$

$$= 10^{10·64}$$

$$10^{·64} = 4·365$$

So that $10^{10·64} = 43{,}650{,}000{,}000$. The amount is more than 40,000 million pounds.

(ii) $\log \dfrac{1}{240} \times 1·005^{6000} = 6000 \log 1·005 - \log 240$

$$= ·00217 \times 6000 - 2·38021$$

$$= 13·02 - 2·38$$

$$= 10·64$$

antilog ·64 = 4·365

Hence: antilog 10·64 = 43,650,000,000

Such calculations are of course merely amusing and fantastic. Logarithms can be used equally well for the calculations made by scientists, engineers, and others. By adding or subtracting two or three logarithms we can make calculations that would otherwise take up considerable time. Suppose we need to find:

$$\sqrt{\frac{8\cdot37}{5\cdot63}}$$

We might divide 8·37 by 5·63 and then find the square root of the quotient. But from a table of logarithms we find:

$$\log 8\cdot37 = \cdot9227$$
$$\log 5\cdot63 = \cdot7505$$

(subtract)   ·1722
(divide by 2)   ·0861

(Look up ·0861 in the table of antilogs.)

antilog ·0861 = 1·219.

Here is the other method set out for comparison:

	1·4867			1·4867 (1·219
563)	837			1
	563	22	·48	
			44	
	2740.			
	2252	241	467	
			241	
	4880			
	4504	2429	22600	
			21861	
	3760			
	3378			
	3820			
	3941			

We are indebted to John Napier, Baron of Merchiston in Scotland, for the invention of logarithms; he first announced his invention in 1614. The logarithms we use, giving numbers as powers of 10, were first calculated by Henry Briggs who was a great friend and admirer of Napier.

The amount of labour involved in calculating the logarithms of the prime numbers was enormous. One of Briggs' methods depends on the fact that in finding the square root of a number we divide the logarithm by 2. We have already used this idea.

$$2 = 10^{.30103}$$

$$\sqrt{2} = 2^{\frac{1}{2}} = 10^{.30103 \times \frac{1}{2}}$$

(That is to find the logarithm of $\sqrt{2}$ we divide the logarithm of 2 by 2.)

In finding the logarithm of 5 Briggs began with 1 and 10, because he knew the logarithms of 1 and 10.

$$1 = 10^0, \text{ so that } \log 1 = 0,$$

$$10 = 10^1, \text{ so that } \log 10 = 1.$$

                                    logarithm
Now write $A = 1$ 0
                 $B = 10$ 1

(Now find the square root of AB, i.e. of 1×10 or simply 10 by the usual process of arithmetic. Call the result C.)

                                    logarithm
$C = \sqrt{AB} = \sqrt{1 \times 10} = 3.162277$ 0.5 (half of $\overline{0+1}$)

(Now find the square root of BC, i.e. of 10×3.162277. Call the result D:

$\log D = $ half of $(\log B + \log C) = \frac{1}{2} (1 + 0.5) = .75$.)

$$D = \sqrt{BC} = \sqrt{31.62277} = 5.623413 \qquad 0.75$$

(We proceed in this way till we find the logarithm of a number which is so close to 5 that the difference does not matter.)

$$E = \sqrt{CD} = 4 \cdot 216964 \qquad 0 \cdot 625$$
$$F = \sqrt{DE} = 4 \cdot 869674 \qquad 0 \cdot 6875$$
$$G = \sqrt{DF} = 5 \cdot 232991 \qquad 0 \cdot 71875$$
$$H = \sqrt{FG} = 5 \cdot 048065 \qquad 0 \cdot 703125$$

We are getting nearer and nearer to 5, sometimes a little above and sometimes a little below, but always nearer at each step. Just when the alphabet begins to run out we reach

$$Z = \sqrt{XY} = 5 \cdot 000000 \qquad 0 \cdot 6989700$$

and 0·6989700 is the logarithm of 5, true to 7 figures.

We now have much easier methods of calculating logarithms, methods that entail much simpler calculations. We begin by calculating the logarithms as powers of a number called e, which is equal to 2·718281828.... It seems a fantastic thing to do, but it happens to be an easy method, because there are series of numbers which can be added up to give this kind of logarithm.

e is itself the sum of a series:

$$e = 1 + 1 + \frac{1}{1 \times 2} + \frac{1}{1 \times 2 \times 3}$$
$$+ \frac{1}{1 \times 2 \times 3 \times 4} + \frac{1}{1 \times 2 \times 3 \times 4 \times 5} + \text{etc.}$$

The series goes on indefinitely, but the numbers in it grow less and less:

$$1 + 1 + \cdot 5 + \cdot 1\dot{6} + \cdot 041\dot{6} + \cdot 008\dot{3} + \text{etc.}$$

so that it is easy to find the value of e to any degree of accuracy.

We can calculate the logarithm of a number a to base e; we will call it x. That is $a = e^x$. We can then find the logarithm of e to base 10; suppose $e = 10^k$. (We shall see later on how this logarithm is found.) Then

$$a = e^x = (10^k)^x = 10^{kx}$$

We find the logarithm of a to base e (that is we find x), and then multiply x by k which is the logarithm of e to base 10. It is known that $\log_{10} e = \cdot43429$. Hence we find the logarithms of numbers to base e, and then multiply them by $\cdot43429$ to change them to base 10. Logarithms to base e are called Naperian logarithms after the inventor.

A useful series for finding logarithms to base e is:

$$\log_e(p+1) - \log_e p$$

$$= 2 \left\{ \frac{1}{2p+1} + \frac{1}{3} \cdot \frac{1}{(2p+1)^3} + \frac{1}{5} \cdot \frac{1}{(2p+1)^5} + \text{etc.} \right\}$$

The series runs on indefinitely, but the numbers rapidly get less, so that we soon reach a point where we can cease to bother about them.

Now let us see how we can use the series. Let us find $\log_e 2$. We put $p = 1$ wherever it comes in the series. This gives us:

$$\log_e 2 - \log_e 1 = 2 \left\{ \frac{1}{3} + \frac{1}{3} \times \frac{1}{3^3} + \frac{1}{5} \times \frac{1}{3^5} + \frac{1}{7} \times \frac{1}{3^7} + \text{etc.} \right\}$$

Now $\log_e 1 = 0$, since $e^0 = 1$. Hence:

$$\log_e 2 = 2 \left\{ \frac{1}{3} + \frac{1}{3} \times \frac{1}{3^3} + \text{etc. as before} \right\}$$

and now we have to find the value of this series. Obviously the terms of the series rapidly get less and less; we soon reach the point at which they can be ignored.

It is a useful plan to write out the powers of 3 first, and then find the appropriate fraction of each. We add the fractions thus obtained, and multiply the sum by 2 (the 2 which is a factor of the series).

$$\frac{1}{3} = \cdot333333 \qquad\qquad \cdot333333$$

(divide by 9) $\quad \dfrac{1}{3^3} = \cdot037037 \qquad (\tfrac{1}{3}) \quad \cdot012346$

$$\frac{1}{3^5} = \cdot004115 \qquad (\tfrac{1}{5}) \quad \cdot000823$$

$$\frac{1}{3^7} = \cdot000457 \qquad (\tfrac{1}{7}) \quad \cdot000065$$

$$\frac{1}{3^9} = \cdot000051 \qquad (\tfrac{1}{9}) \quad \cdot000006$$

$$\frac{1}{3^{11}} = \cdot000006 \qquad (\tfrac{1}{11}) \quad \cdot000001$$

$$\overline{\qquad\qquad\qquad \cdot346574}$$

$$\cdot346574 \times 2 = \cdot693148$$

$\cdot69315 = \log_e 2$. We multiply this number by $\cdot43429$, and thus find $\log_{10} 2$:

$$\begin{array}{r}
\cdot69315 \\
\cdot43429 \\
\hline
\cdot277260 \\
\cdot020794 \\
\cdot002773 \\
\cdot000139 \\
\cdot000062 \\
\hline
\cdot301028
\end{array}$$

Ignoring the last two figures we find that $\log_{10} 2 = \cdot3010$. If we wish to find log 2 more accurately (say to 7 figures) we should have to extend the calculation. The value of each term would have to be worked out to nine figures, and we should have to sum several more terms in order to ensure that the value of those ignored was not so great as 000000001. Also we should need a more accurate value for $\log_{10} e$ (a very accurate value is 0·4342944819).

There is a point that needs to be cleared up: how is it possible to find $\log_{10} e$? We can find $\log_e 10$ easily enough. We have shown how to find $\log_e 2$ from the series. To find $\log_e 3$ we put $p = 2$ wherever it occurs in the series. Thus we get:

$$\log_e 3 - \log_e 2 = 2 \left\{ \frac{1}{5} + \frac{1}{3} \times \frac{1}{5^3} + \frac{1}{5} \times \frac{1}{5^5} + \text{etc.} \right\}$$

We sum the series as before—it is much easier this time—and thus obtain $\log_e 3 - \log_e 2$. We have already found $\log_e 2$ ($= \cdot 69315$), so we merely have to add this.

Now let us find $\log_e 10$. We write $p = 9$ in the series:

$$\log_e 10 - \log_e 9 = 2 \left\{ \frac{1}{19} + \frac{1}{3} \times \frac{1}{19^3} + \frac{1}{5} \times \frac{1}{19^5} + \text{etc.} \right\}$$

$\log_e 9$ is of course equal to $2 \log_e 3$. So we sum the series and add twice $\log_e 3$.

Thus we find $\log_e 10$, and we want to find $\log_{10} e$. Suppose $\log_e 10 = a$. (Simply call it a for convenience.) Then $10 = e^a$; that is what the logarithm is—the power to which e must be raised to equal 10:

$$10 = e^a$$

Now take the ath root of each side:

$$10^{\frac{1}{a}} = e$$

That is to say $\log_{10} e = \frac{1}{a}$. $\log_{10} e$ is the reciprocal of $\log_e 10$.

That is of course true for any number, because we can write any number in place of 10 without affecting the argument.

The method of finding logarithms to base 10 which has just been described is incomparably easier than that used by Briggs in his original computations of a table of logarithms. His calculation of the logarithm of 5 involved the extraction of 24 square roots. We can find it by noting that:

$$\log 5 = \log \tfrac{10}{2} = \log 10 - \log 2$$
$$= 1 - \cdot 30103$$
$$= \cdot 69897$$

The calculation from the series is given here for comparison with Briggs' method:

$$\log_e 5 - 2 \log_e 2 = 2 \left\{ \frac{1}{9} + \frac{1}{3} \times \frac{1}{9^3} + \frac{1}{5} \times \frac{1}{9^5} + \text{etc.} \right\}$$

$1/9 = \cdot 111111$		$\cdot 111111$
$1/9^3 = \cdot 001372$	$(\tfrac{1}{3})$	$\cdot 000457$
$1/9^5 = \cdot 000017$	$(\tfrac{1}{5})$	$\cdot 000003$
$1/9^7 = \cdot 000000$		

$$\cdot 111571$$
$$2$$

Add $2 \log_e 2 \quad \cdot 223142$
$$1 \cdot 386294$$
$$1 \cdot 609436$$
$$\cdot 43429$$
$$\cdot 643774$$
$$\cdot 048283$$
$$\cdot 006438$$
$$\cdot 000322$$
$$\cdot 000145$$
$$\cdot 698962$$

$\log_{10} 5 = \cdot 69896$, with an error of 1 in the last place. To four figures $\log_{10} 5 = \cdot 6990$.

There is another way in which logarithms can be employed with great advantage. Sometimes we wish to draw a diagram illustrating, say, a range of wave-lengths. Wavelengths vary from many metres down to the most minute fractions of metres. Now it happens that our interest is as much concerned with the minute waves as with the very long waves. Gamma rays (emitted by radioactive substances) have a wave-length of about $\cdot 0000000001$ cm.; X-rays vary from about $\cdot 0000005$ to $\cdot 000000001$ cm. Ultra-violet rays are longer still, then come visible rays, the infra-red rays, and finally the long wireless waves running up to thousands of metres in length.

If we were to make a scale drawing representing 10^5 cm., 10^4 cm., 10^3 cm., 10^2 cm., 10 cm., 1 cm., 10^{-1} cm., 10^{-2} cm., etc., the first part of the diagram—10^5 cm.–10^4 cm.—would occupy 90,000 units of our scale, whereas everything lower than 1 cm. would be crushed up into a small fraction of a unit. We want to show the tiny gamma rays on the diagram just as clearly as the long wireless rays. We get over the difficulty by using a logarithmic scale. Instead of using the numbers:

$$10^5 \quad 10^4 \quad 10^3 \quad 10^2 \quad 10 \quad 1 \quad 10^{-1} \quad 10^{-2} \quad \text{etc.}$$

we use their logarithms:

$$5 \quad 4 \quad 3 \quad 2 \quad 1 \quad 0 \quad -1 \quad -2 \quad \text{etc.}$$

These numbers are equally spaced along the diagram.

Before leaving this subject let us see how it has been developed. We began with the quite simple ideas of the addition and subtraction of indices in multiplying and dividing. Keeping the rule universal we found meanings for

10^0, 10^{-1}, 10^{-2}, etc. that fitted in with the rule. We proceeded to find meanings for fractional indices—again these had to fit in with the rule. We then saw how fractional indices could be made into a table of logarithms for use in calculations. And finally we saw how logarithmic scales can be used.

The whole subject is an interesting example of arithmetical logic—beginning with a simple and obvious idea and developing this along the lines of continuity and universality.

IV

Proportion

THE idea of proportion is as fundamental to arithmetic as
are the ideas of continuity and universality. The idea of
proportion is so simple, and we assume it so readily, that it
is not easy to define what it is.

We say that two quantities are in proportion when one is
always the same number of times the other. If one kind of
quantity is always twice the other, or always 3 times the
other, or always 5¼ times the other, and so on.

We are given the cost of 1 article and asked to find the
price of 9 articles. We assume that the total cost is proportional to the number of articles—9 articles cost 9 times
as much as 1 article. If that were not so we could not even
begin to find an answer. If there were, for example, a
reduction for quantities we should want to know more about
the sum.

"6d. each, 5s. 6d. per dozen, £2. 2s. per 100," says a
catalogue of plants, and it adds "not less than 6 at the dozen
rate; not less than 50 at the 100 rate." There are evidently
three different proportions: for 1 to 5 plants, for 6 to 49
plants, and for 50 to 100 plants. Whatever calculations we
make we assume one or other of these proportions. Thus
4 dozen plants at the dozen rate cost 22s., and we realize
that it would be better to have 50 plants at the 100 rate for
21s. But, according to the stated terms, we must have one
proportion or another.

If the speed of a train is constant then the distance it travels is proportional to the time. In 2 hours it travels twice the distance it travels in 1 hour. When we find that light takes 500 seconds to reach us from the sun we accept the fact that the speed of 186,000 miles per second is constant and we assume that the time is proportional to the distance —double the distance and we double the time.

In these examples it may be seen that in each we have one constant quantity, and two variable quantities that are in proportion. The cost of a single article is constant; the number of articles and the total cost are in proportion. The speed is constant; the time and distance are in proportion.

There is a simple and convenient way of expressing this in algebra. Call the two variable quantities x and y, and call the constant quantity c (c is always the same in any particular case of proportion). Then

$$x = cy,$$

that is to say the quantity x is c times the quantity y.

Take a simple example: if articles are sold each at the same price, then the total cost of a number of articles is proportional to the number of articles:

$$\text{total cost} = c \times \text{number of articles.}$$

Here c obviously stands for "the cost of one article"; this is a constant quantity for any particular example, but it would be different in another example.

$$\text{Distance in miles} = c \times \text{time in seconds.}$$

For the distance travelled by light c stands for "186,000 miles per second".

The production of machines is usually proportional to the

time they are working. Thus: a machine produces 9000 bolts
in $7\frac{1}{2}$ hours:

$$\text{number of bolts} = c \times \text{number of hours};$$

$$\therefore \quad c = \frac{\text{number of bolts}}{\text{number of hours}}$$

$$= \frac{9000}{7\frac{1}{2}}$$

$$= 1200 \text{ bolts per hour.}$$

We can use the value of c to find the number of bolts made
in any number of hours.

"Rule of Three" sums are particular examples of this
kind of proportion. In everyday life the two parts of such
sums are usually done at widely different times and often
by different people. The selling price of envelopes may be
decided by dividing the total cost (including profits) by the
total number of envelopes; that may be done at the factory.
The buyer of envelopes and the shopkeeper are concerned
with the other half of the problem—to find what must be
paid for 2, 3 or more packets of envelopes.

A research worker may find the coefficient of expansion of
a certain steel, by division. The coefficient thus found may
be used by himself or by others to find the total expansion
of a certain length of steel. Indeed coefficients of expansion
are printed in reference books so as to facilitate the working
of the second part of the sum.

The kind of proportion we have been discussing is simple
direct proportion. It is not the only kind of proportion.

One quantity may be directly proportional to the square
of another quantity:

$$x = cy^2$$

that is, x is proportional to y^2.

The area of a square is proportional to the square of the length of a side:

$$\text{area of square} = c \times \text{side}^2$$

Now a square with sides 1 inch long has an area of 1 square inch:

$$1 = c \times 1^2$$

so that $\qquad c = 1.$

Indeed the unit of area (the square inch) is chosen so that c may equal 1 when the lengths are in inches. We now have:

$$\text{area of square} = \text{side}^2$$

The area of a square is also proportional to the square of any other line in it, provided we take the same kind of line for each square we consider. For any particular kind of line c will have the same value for every square; but if we switch over to another kind of line then c will have a different value.

Suppose we take the diagonal, which is the same sort of line for every square. Then:

$$\text{area of square} = c \times \text{diagonal}^2,$$

and we want to find c.

Look at this 1 inch square. ABD is a right-angled triangle, so that:

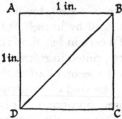

$$BD^2 = AB^2 + AD^2$$
$$= 1^2 + 1^2$$
$$= 2$$

In a 1 inch square the square of the diagonal is 2 square inches.

$$\text{Area} = c \times 2.$$

(The area is 1 square inch.)

$$\therefore \quad 1 = c \times 2,$$
$$\text{or} \quad c = \tfrac{1}{2}.$$

Hence in any square:

$$\text{area} = \tfrac{1}{2}\ \text{diagonal}^2$$

As a further example let us take a line from a corner of
a square to the middle of one of
the opposite sides. ABC is a right-
angled triangle, and so:

$$AB^2 = AC^2 + BC^2$$
$$= 1^2 + (\tfrac{1}{2})^2$$
$$= 1\tfrac{1}{4}\ (\text{or } \tfrac{5}{4})$$

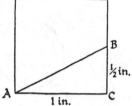

In a 1 inch square:

$$\text{area} = c \times AB^2,$$
$$1 = c \times 1\tfrac{1}{4}$$
$$\therefore\quad c = \frac{1}{1\tfrac{1}{4}}$$
$$= \tfrac{4}{5}$$

Hence in any square:

area $= \tfrac{4}{5} \times$ square of line from a corner to the middle of an
opposite side.

The reason we can be sure of all squares is that all squares
are similar figures. Any square is a magnified version of any
smaller square.

$XYZW$ is a magnified version of $ABCD$. $XY = 1\tfrac{1}{4}AB$,

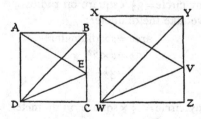

$YW = 1\frac{1}{4}BD$, $WV = 1\frac{1}{4}DE$, $XV = 1\frac{1}{4}AE$, and so on with any other similar lines. The area of a square is proportional to the square of any of these lines, provided we take the same kind of line in each square we consider.

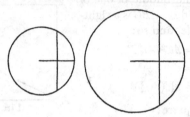

Circles also are similar figures; any circle is a magnified version of any smaller circle. Hence:

area of circle $= c \times$ square on any line,

provided the same kind of line is taken in each circle considered—radius, diameter, a line at right angles to a radius at its mid-point, etc.

It is a fact, which can be demonstrated in various ways, that a circle of radius 1 inch has an area of about $3\frac{1}{7}$ square inches:

$$\text{area} = c \times \text{radius}^2$$
$$3\frac{1}{7} = c \times 1^2$$
$$c = 3\frac{1}{7}$$

∴ area of any circle $= 3\frac{1}{7} \times$ square on radius.

Suppose we take the diameter:

$$\text{area} = c \times \text{diameter}^2$$
$$3\frac{1}{7} = c \times 2^2$$
$$c = 3\frac{1}{7} \div 4$$
$$= \tfrac{11}{14}$$

∴ area of any circle $= \tfrac{11}{14} \times$ square on diameter.

The radius of the earth at the equator is about 3963 miles and that of the moon 1080 miles, so that the earth is $\frac{3963}{1080} = 3 \cdot 67$ or about $3\frac{2}{3}$ times as wide as the moon. The area of the earth is therefore $(3\frac{2}{3})^2$ times as great as the area of the moon:

$$(3\tfrac{2}{3})^2 = (\tfrac{11}{3})^2 = \tfrac{121}{9}$$
$$= 13\tfrac{4}{9}$$

So that the surface of the earth is about $13\frac{1}{2}$ times as great as that of the moon. The "full earth" as seen from the moon would appear $13\frac{1}{2}$ times as big as the full moon seen from the earth.

The radius of the sun is 433,000 miles, so that the sun is $\frac{433000}{3963}$ = about 109 times as wide as the earth. The area of the sun is therefore 109^2 times the area of the earth, or about 12,000 times as great.

Squares and circles are two examples of the general proposition that the areas of similar figures are proportional to the squares of similar lines in these figures.

Equilateral triangles are similar figures; any equilateral triangle is a magnified version of any smaller equilateral triangle. Regular hexagons are similar to each other, and so with other regular figures.

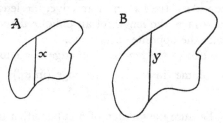

To take the most general case of all let us think of an irregular shape A. B is a magnified version of A, and therefore of similar shape. y is a line in B similar to x in A; when A is magnified to B then x is magnified to y:

$$\text{area } A = cx^2,$$
$$\text{area } B = cy^2,$$

and c is the same for both figures.

We can get rid of c by dividing:

$$\frac{\text{area } A}{\text{area } B} = \frac{cx^2}{cy^2} = \frac{x^2}{y^2}$$

Suppose that y is 1·7 times x, ($y = 1·7x$), then:

$$\frac{\text{area } A}{\text{area } B} = \frac{x^2}{(1·7x)^2} = \frac{1}{1·7^2} = \frac{1}{2·89}$$

Hence area B is 2·89 times as great as area A, when a line in B is 1·7 times a similar line in A.

We may have one quantity proportional to the cube of another quantity:

$$x = cy^3$$

that is, x is proportional to y^3.

The volume of a cube is proportional to the cube of any line in it, provided we take the same kind of line in every cube we consider—the length of an edge, the length of the diagonal of a face, the length of an oblique diagonal from one corner to the opposite corner, etc.

When we choose the length of an edge of a cube:

$$\text{volume (in cu. in.)} = c \times \text{edge (in in.)}^3$$
$$= \text{edge}^3$$

since $c = 1$, because the volume of a cube with 1 inch edges is 1 cubic inch.

The volume of a sphere is proportional to the cube of the

length of the radius, the diameter, or any other line, provided
we take the same kind of line in every sphere we consider.

$$\text{Volume of sphere} = c \text{ radius}^3.$$

It can be shown that the volume of a sphere of radius 1 inch
is $\frac{4}{3} \times 3\frac{1}{7}$, so that
$$\frac{4}{3} \times 3\frac{1}{7} = c \times 1^3$$
or
$$c = \frac{4}{3} \times 3\frac{1}{7}$$
$$\text{Volume of sphere} = \frac{4}{3} \times 3\frac{1}{7}r^3$$

We found that the radius of the earth is about $3\frac{2}{3}$ times that
of the moon. Hence the volume of the earth is $(3\frac{2}{3})^3$ times
that of the moon:
$$(3\tfrac{2}{3})^3 = (\tfrac{11}{3})^3 = \tfrac{1331}{27}$$
$$= \text{about } 49,$$

so that the volume of the earth is about 49 times as great as
that of the moon.

We also found that the radius of the sun is about 109
times as great as that of the earth. Hence the volume of the
sun is 109^3 times as great as that of the earth:
$$109^3 = \text{nearly } 1,300,000,$$

so that the sun has nearly 1,300,000 times the volume of the
earth.

Cubes and spheres are particular cases of similar solids.
The volumes of any similar solids are proportional to the
cubes of similar lines, provided we take the same kind of line
in each of the similar solids we are considering.

In *The First Men in the Moon* Mr Wells writes: "I had
forgotten that on the moon, with only an eighth part of the
earth's mass and a quarter of its diameter, my weight was
barely a sixth of what it was on earth. But now that fact
insisted on being remembered."

One of the facts apparently did not insist on being remembered correctly.

We found that the earth has about $3\frac{2}{3}$ times the radius (or equally the diameter) of the moon. That is the moon has $\frac{3}{11}$ the radius of the earth; so that a quarter is a reasonable rough approximation.

The volume of the earth is about 49 times as great as that of the moon, so that the moon has $\frac{1}{49}$ of the volume of the earth. But the density of the moon is only about $\frac{3}{5}$ of that of the earth; so that the mass of the moon is only about $\frac{3}{5} \times \frac{1}{49}$ of the earth's mass:

$$\frac{3}{5} \times \frac{1}{49} = \text{about } \frac{1}{81\frac{2}{3}}$$

"An eighth part of the mass" should therefore be an eightieth. It is odd that the error should be repeated in edition after edition of the book; but that is far from uncommon with arithmetical errors.

Look again at a quantity proportional to the square of another:

$$x = cy^2$$

If we take the square root of each side we have:

$$\sqrt{x} = \sqrt{c}\, y$$

or

$$y = \frac{1}{\sqrt{c}} \sqrt{x}$$

Now $\frac{1}{\sqrt{c}}$ is just as much a constant as c itself. $\left(2, \sqrt{2}, \frac{1}{\sqrt{2}} \text{ are}\right.$ all constants—each stands for a number.$\left.\vphantom{\frac{1}{\sqrt{2}}}\right)$ Let us write k for $\frac{1}{\sqrt{c}}$:

$$y = k\sqrt{x}$$

that is y is proportional to the square root of x. This is only another way of saying that x is proportional to the square of y.

We might say that the length of a pendulum (call the length l) is proportional to the square of the time it takes to make a swing: $$l = ct^2$$

That is true enough but it is not a good way of expressing it, because we want to be able to measure a pendulum and thus find its time of swing. It is equally true that:

$$\sqrt{l} = \sqrt{c}\, t,$$

or $$t = k \sqrt{l} \quad \left(\text{where } k = \frac{1}{\sqrt{c}}\right).$$

That is, the time of swing is proportional to the square root of the length. If we make the length of a pendulum four times as great we double the time of swing. When a single swing is considered the constant turns out to be $\dfrac{\pi}{\sqrt{g}}$. ($\pi = 3\frac{1}{7}$, $g = 32$ ft. per sec. per sec.) Hence in every case:

$$t = \frac{\pi}{\sqrt{g}} \sqrt{l} \quad \text{or} \quad \pi \sqrt{\frac{l}{g}}$$

INVERSE PROPORTION

So far we have been dealing with direct proportion. A quantity may also be in proportion to the inverse of another, not to y but to $\dfrac{1}{y}$:

$$x = \frac{c}{y}$$

If we multiply both sides by y we get

$$xy = c,$$

which is another way of saying the same thing.

The time taken by a number of men to do a certain amount of mechanical work is often inversely proportional to the number of men. Double the number of men and we halve the time. There are numerous exceptions: it does not follow that because 1 man takes a minute, 60 men will take a second. We have to use common sense in arithmetical calculations.

It is more exactly true of machines that the time taken is inversely proportional to the number of machines. Indeed the inverse proportion only applies to the kind of work that could be done by machines. It might apply for example to digging trenches or laying bricks or to the kind of arithmetic that can be done on number machines. But it would not apply to problems that demand original thought. Because a man can write a sonnet in ten hours it does not follow that ten men could write one in one hour. Though something like that has been attempted at Oxford.

9 machines produce a certain number of screws in 6 hours. We can use the inverse proportion equation with some confidence:

$$\text{time} = \frac{c}{\text{number of machines}}$$

$$6 = \frac{c}{9},$$

or $$c = 54.$$

Hence: $$\text{time} = \frac{54}{\text{number of machines}}$$

6 machines would take $\frac{54}{6} = 9$ hours; 1 machine would take $\frac{54}{1} = 54$ hours; 2 machines would take $\frac{54}{2} = 27$ hours, and so on.

A well-known case of inverse proportion is the relation between the volume and pressure of a mass of gas. We have to see that nothing varies but the volume and the pressure. We must see that no gas escapes; if the gas is heated when we compress it we must give it time to cool down. If we take these precautions we shall find that by doubling the pressure we halve the volume of the gas. We have the proportion:

$$\text{volume of gas} = \frac{c}{\text{pressure of gas}}$$

or $$V = \frac{c}{P}$$

We can write this: $VP = c.$

This is the law known as Boyle's Law: the volume of any mass of gas multiplied by the pressure of the gas is constant —provided the temperature does not change.

In the equation c is constant for any particular mass of gas; but it would be different for another mass of gas. If we had a million cubic feet at atmospheric pressure in one container, and one cubic foot at atmospheric pressure in another, the constant would be a million times as great for the first mass of gas as for the second.

Here is an interesting and important sort of proportion:

O is a bright light—as small as possible, so that it may throw clear sharp shadows. A is a square of cardboard. It throws a square shadow on a screen at B, and larger shadows when the screen is moved to C or D.

B is twice as far from O as A is; the sides of the square at B are twice as great as those of A, and so the area of B is $2^2 = 4$ times as great. The sides of the square at C are three times as great as those of A, and the area is $3^2 = 9$ times as great. The square at D is $4^2 = 16$ times as great. Thus the

area of a square is proportional to the square of its distance
from O.

Area $= c \times$ square of distance from O.

Now think of the amount of light that would fall on each
of these squares if the others were out of the way. At A the

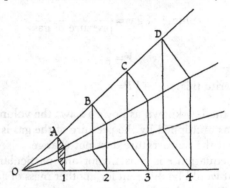

light is concentrated on a small square. At B the same
amount of light is spread over a space four times as great
as A; so that the illumination is only a quarter as great.
At C the light (still the same amount of light) is spread over
a space nine times as great as A; the illumination is only a
ninth as great. And at D the illumination is only a sixteenth
as great.

The illumination is proportional, not to the square of the
distance, but to its inverse.

$$x \text{ (illumination)} = \frac{c}{y^2} \ (y = \text{distance from light}).$$

$$x = \frac{c}{y^2}$$

is sometimes called "the inverse square law".

That is why the light of a lamp fades away so rapidly in the open. At a distance of 10 feet, for example, the illumination is only one-hundredth $\left(\dfrac{1}{10^2}\right)$ of what it is at 1 foot. In a room there is much reflection from walls and ceiling (unless these are very dark) and the illumination is more even.

When a fire has just been lighted in a room its effect dies away very rapidly with distance—double the distance and we get only a quarter of the heat. After a time the walls and furniture become warm and these also radiate heat.

The heat and light which the sun gives out to the planets are not much affected by reflection or radiation from other objects, because space is almost entirely empty. A little light is reflected to us from the moon, but that is almost all. The amounts of heat and light which any spot receives from the sun are almost exactly proportional to the inverse of the square of the distance of the spot from the sun. Suppose we consider the amounts of heat and light received by Mercury (the planet nearest the sun), the earth, and Neptune (the outermost planet but one—Pluto). We shall have to think of an equal area on each planet—say a square yard or a square mile.

The distances from the sun in millions of miles are: Mercury 36, earth 93, Neptune 2794. In every case:

x (amount of heat and light received—per square mile, say)

$=\dfrac{c}{y^2}$ (y = distance from source of light and heat—the sun)

$$x=\frac{c}{y^2}$$

We can get rid of c by dividing, thus:

$$\frac{\text{light and heat on Mercury}}{\text{light and heat on earth}}$$

$$= \frac{c}{(\text{Mercury's distance})^2} \div \frac{c}{(\text{earth's distance})^2}$$

$$= \frac{c}{36^2} \div \frac{c}{93^2}$$

$$= \frac{93^2}{36^2}$$

$$= \text{about } 6 \cdot 7$$

That is, any patch of ground on Mercury receives nearly seven times as much heat and light from the sun as a corresponding patch on the earth—a really terrific glare and scorching heat.

$$\frac{\text{Light and heat on Neptune}}{\text{Light and heat on earth}} = \frac{c}{2794^2} \div \frac{c}{93^2}$$

$$= \frac{93^2}{2794^2}$$

$$= \text{about } \frac{1}{900}$$

That is, a patch of ground on Neptune receives only $\frac{1}{900}$ part of the light and heat received by a corresponding patch on the earth. Neptune must be a dark and chilly planet.

The constant c which comes into the equation is the same for the whole solar system, because the source of light and heat (the sun) is the same for all. But if the sun were to give out twice as much light and heat the constant would be doubled—it would still be the same for all.

More elaborate proportions are possible than any we have yet considered. We may have, for example, a quantity proportional not merely to one other quantity but to several.

According to Newton's Law of Gravitation there is an attraction between any two bodies—each attracts the other, and the attraction of A for B is the same as the attraction of B for A. The attraction between two bodies is proportional to three quantities: it is directly proportional to the mass of each of the two bodies, and it is inversely proportional to the square of the distance between their centres of gravity. As two bodies move apart the attraction that each has for the other dies away at the same rate that heat and light die away.

We can write the proportion as an equation:

attraction between two bodies

$$= G \frac{\text{mass of one body} \times \text{mass of other body}}{(\text{distance})^2}$$

G stands for the constant in the proportion; it is called the constant of gravitation. This constant is extraordinary; it appears to be a universal constant—the same for the attraction between any two bodies anywhere. It is not like the constant for the radiation of heat and light from the sun; this constant is the same for the whole solar system, but it would be a different constant if we had a different sun, or if the sun's energy were to increase or fade away.

We can write the equation for Newton's law more simply by putting M_1 and M_2 for the masses of the two bodies and d for the distance apart of their centres of gravity. The equation then becomes:

$$\text{attraction} = G \frac{M_1 M_2}{d^2}$$

We can find the value of G if we remember that the pull of the earth on a mass of 1 pound on the earth's surface is a

weight of 1 pound (that is what the weight is). We write
these quantities in the equation:

$$\text{1 lb. weight} = G\,\frac{\text{1 lb. mass} \times \text{mass of earth}}{(\text{radius of earth})^2}$$

We now have to find the mass of the earth. The volume of
a sphere is $\frac{4}{3} \times 3\frac{1}{7} \times r^3$ (where r is the radius of the sphere).
The earth is so nearly spherical that we can treat it as a
sphere.

The volume of the earth is:

$$\frac{4}{3} \times 3\frac{1}{7} \times 4000^3 \text{ cubic miles}$$
$$= \frac{4}{3} \times 3\frac{1}{7} \times 4000^3 \times 5280^3 \text{ cubic feet}$$

Now a cubic foot of water weighs $62\frac{1}{2}$ lb. and the earth as
a whole is $5\frac{1}{2}$ times as dense as water. Hence the average
density of the earth is $62\frac{1}{2} \times 5\frac{1}{2}$ lb. per cubic foot.

The whole mass of the earth is therefore:

$$\frac{4}{3} \times \frac{22}{7} \times 4000^3 \times 5280^3 \times 62\frac{1}{2} \times 5\frac{1}{2} \text{ lb.}$$

The attraction equation:

$$\text{1 lb. weight} = G\,\frac{\text{1 lb. mass} \times \text{mass of earth}}{(\text{radius})^2}$$

becomes:

$$\text{1 lb. weight} = G\,\frac{\text{1 lb.} \times \frac{4}{3} \times \frac{22}{7} \times 4000^3 \times 5280^3 \times 62\frac{1}{2} \times 5\frac{1}{2} \text{ lb.}}{4000^2 \times 5280^2}$$

(The factor 5280^2 in the denominator changes the miles to
feet, to correspond with feet in the numerator. The units
throughout are feet and pounds.)

So that
$$G = \frac{1 \times 4000^2 \times 5280^2}{1 \times \frac{4}{3} \times \frac{22}{7} \times 4000^3 \times 5280^3 \times 62\frac{1}{2} \times 5\frac{1}{2}}$$
$$= \frac{3 \times 7 \times 2 \times 2}{4 \times 22 \times 4000 \times 5280 \times 125 \times 11}$$
$$= (\text{after a little division}) \; 3 \cdot 3 \times 10^{-11}$$

Set out in full this would be ·000000000033.

G is a very small number, as one might expect when one considers that the attraction of the whole earth on a mass of 1 pound at the earth's surface is only 1 pound weight. Equally the attraction of a mass of 1 pound at the earth's surface on the whole earth is 1 pound weight.

Let us see how we can use the gravitational constant. What is the attraction between two ships each weighing 20,000 tons when the distance between their centres of gravity is 100 feet?

We must keep to the same units throughout, pounds and feet; otherwise the value of the constant would be different. Thus 20,000 tons = 20,000 × 2240 lb. Since the weights of the two ships are equal we have M^2 instead of $M_1 M_2$.

$$\text{Attraction in pounds weight} = G\,\frac{M^2}{d^2}$$

$$= \frac{3\cdot3 \times (20{,}000 \times 2240)^2}{10^{11} \times 100^2}$$

$$= \frac{3\cdot3 \times 2^2 \times 10^8 \times 2\cdot24^2 \times 10^6}{10^{11} \times 10^4}$$

$$= 3\cdot3 \times 4 \times 2\cdot24^2 \times 10^{-1}$$

$$= \text{about } 6\cdot6 \text{ lb. weight}$$

At a distance of 200 feet the attraction would be a quarter of that. Everything would be the same in the equation except that 200^2 would be written in place of 100^2 in the denominator. This would make the denominator four times as great, and therefore it would reduce the number to a quarter of its previous value, i.e. to a little more than $1\frac{1}{2}$ lb. (We double the distance and quarter the attraction.) At a distance of 5000 feet—say a mile—the distance would be

50 times as great as at 100 feet, and therefore the attraction would be $\dfrac{1}{50^2} = \dfrac{1}{2500}$ times as great.

$$\dfrac{6 \cdot 6}{2500} \text{ lb.} = \dfrac{26 \cdot 4}{10{,}000} \text{ lb.}$$
$$= \cdot 00264 \text{ lb.}$$
$$= \cdot 04 \text{ oz.,}$$

that is about a twenty-fifth of an ounce. When they are a mile apart each ship exerts on the 20,000 ton mass of the other a pull equal to the twenty-fifth part of an ounce.

And now let us see what a 10 stone man would weigh on the moon.

Attraction in pounds weight $= G \dfrac{140 \text{ lb.} \times \text{mass of moon}}{(\text{radius of moon})^2}$.

The radius of the moon is 1080 miles. So that the volume is $\frac{4}{3} \times 3\frac{1}{7} \times 1080^3$ cubic miles $= \frac{4}{3} \times \frac{22}{7} \times 1080^3 \times 5280^3$ cubic feet. The density of the moon is $\frac{3}{5}$ that of the earth, and the density of the earth is $5\frac{1}{2}$ times that of water. So that the density of the moon is:

$$\tfrac{3}{5} \times 5\tfrac{1}{2} \times 62\tfrac{1}{2} \text{ lb. per cubic foot.}$$

This gives the mass of the moon as:

$$\tfrac{4}{3} \times \tfrac{22}{7} \times 1080^3 \times 5280^3 \times \tfrac{3}{5} \times 5\tfrac{1}{2} \times 62\tfrac{1}{2} \text{ lb.}$$

We fill in this value in the equation, and also the radius of the moon. This gives:

attraction in pounds weight

$$= \dfrac{3 \cdot 3 \times 140 \times \tfrac{4}{3} \times \tfrac{22}{7} \times 1080^3 \times 5280^3 \times \tfrac{3}{5} \times 5\tfrac{1}{2} \times 62\tfrac{1}{2}}{10^{11} \times 1080^2 \times 5280^2}$$

$$= \dfrac{3 \cdot 3 \times 140 \times 4 \times 22 \times 1080 \times 5280 \times 3 \times 11 \times 125}{10^{11} \times 3 \times 7 \times 5 \times 2 \times 2}$$

$=$ (after some more cancelling and multiplying and dividing) 22·8

$=$ about 23 lb.,

that is, near enough, one-sixth of the man's 10 stone weight on earth.

We can get quite a number of interesting results without using the value we have calculated for G.

Tables often give the masses of sun, moon, and planets, taking the earth's mass as 1. This is usually convenient; it saves the trouble of writing very large numbers. In the following table both masses and radii are given in this form:

	Mass	Radius
Sun	329,000	109
Moon	0·012	0·27
Mercury	0·34	0·35
Venus	0·82	0·96
Earth	1	1
Mars	0·11	0·53
Jupiter	314·5	11
Saturn	94·1	9·6
Uranus	14·4	3·9
Neptune	16·7	4·2

We can use these numbers to find in a very simple way what a 10 stone man would weigh on the sun, moon or any of the planets. We will begin with the sun; we have of course to think of a body of the size and mass of the sun on which it would be possible to exist.

On the sun we have:

$$\text{weight of man in pounds} = G\,\frac{140\text{ lb.} \times 329,000}{109^2}$$

(Since we are going to compare earth and sun we can use the mass 329,000. The factors which would reduce this to pounds are the same for the earth and therefore will cancel out. This is also true of the radius of the sun.)

On the earth we have:

$$140 \text{ lb. weight} = G \, \frac{140 \text{ lb.} \times 1}{1^2}$$

$$\frac{\text{weight on sun}}{140 \text{ lb. weight}} = \frac{G \, \dfrac{140 \times 329{,}000}{109^2}}{G \times 140}$$

$$= \frac{329{,}000}{109^2}$$

$$= \text{about } 28$$

So that our 10 stone man would weigh about 28 times as much on the sun, that is 280 stones or 35 hundredweights.

We could work out the weights on any of the planets in the same way. We will however transport our man to the moon, to show how much easier it is to work out his weight on the moon by this method:

$$\frac{\text{weight on moon}}{140 \text{ lb. weight}} = \frac{G \, \dfrac{140 \times \cdot012}{\cdot27^2}}{G \times 140}$$

$$= \frac{\cdot012}{\cdot27^2}$$

$$= \text{nearly } \cdot165,$$

which is near enough to $\frac{1}{6}$ of his weight on earth—about 23 pounds.

On Mr Wells' moon, it may be remembered, the mass of the moon is one-eighth of that of the earth, and the diameter one-quarter.

$$\frac{\text{Weight on Mr Wells' moon}}{140 \text{ lb. weight}} = \frac{G \, \dfrac{140 \times \frac{1}{8}}{(\frac{1}{4})^2}}{G \times 140}$$

$$= \frac{\frac{1}{8}}{(\frac{1}{4})^2}$$

$$= 2$$

So that a 10 stone man would weigh 20 stones on Mr
Wells' moon, and would therefore be incapable of the long
fantastic jumps that Mr Wells credits him with. It just
shows how careful creators have to be. A creator may have
the most admirable intentions; he may intend his creatures
to be light and airy and even fantastically happy. But a
slight error in his calculations may upset the whole utopia
and bind his creatures more tightly to earth than they are
bound in the world from which he would release them.
A creator has to be an expert mathematician.

Another case of an unusual proportion is to be found in
the solar system. According to Kepler's third law: the
squares of the periodic times of the planets are proportional
to the cubes of the distances of the planets from the sun:

$$(\text{periodic time})^2 = c \ (\text{distance from sun})^3$$

c is a constant for the solar system. There would be a
different constant for another stellar system.

The periodic times are the times the planets take to
revolve round the sun. Taking the earth's year as unit the
periodic times of the planets are:

Mercury	Venus	Earth	Mars	Jupiter	Saturn	Uranus	Neptune
·2408	·6154	1	1·881	11·86	29·46	83·74	164·78

Look again at the equation which gives the proportion.
From this we readily find:

$$c = \frac{(\text{periodic time})^2}{(\text{distance from sun})^3}$$

We can use the earth's distance from the sun as the unit.
With this unit the distances are:

Mercury	Venus	Earth	Mars	Jupiter	Saturn	Uranus	Neptune
·387	·723	1	1·52	5·20	9·54	19·19	30·07

For Mercury:

$$c = \frac{\cdot 2408^2}{\cdot 387^3} = 1 \cdot 000 \text{ (to 3 places)}.$$

For Venus: $c = \frac{\cdot 6154^2}{\cdot 723^3} = 1 \cdot 002.$

For the earth: $c = \frac{1^2}{1^3} = 1.$

The reason why c comes out as 1 is that the periodic time and radius of one of the planets (the earth) have been chosen as units. Since $c = 1$ for this planet it must equal 1 for the other planets. An allowance must be made for inaccuracies in the approximations to the periodic times and the radii.

If other units had been chosen the value of c would be different, but it would still be the same for every planet.

V

Comparisons

I F anyone thinks he has no use for arithmetic let him consider the number of comparisons he makes every day. Most comparisons involve the working of some sort of sum; even when we compare colours we are apt to make some kind of computation of the proportions of the primary colours in them. We are always making comparisons, well or ill, accurately or inaccurately; one of the virtues of arithmetic is that it enables one to make comparisons neatly and accurately. One is proud of having a sense of proportion, and proportion is essentially mathematical—it is a comparison of sizes measured in some way or other. Even where arithmetic does not enter directly and obviously into a comparison a knowledge of the possibilities of arithmetical comparison is not amiss; it does at least rule out many crudities and absurdities.

I saw some time ago the statement that if all the herrings landed at Yarmouth in a season were placed head to tail in a line they would stretch nine times round the world. The writer had evidently been impressed by the vast number of herrings in the catch, and he had looked for a comparison that would show the immensity of the number of herrings. As we have already seen, if we want to make the most of a large number of objects, we place them in a line. We get the greatest possible length by placing the herrings head to tail, and that is what the writer had done.

The idea of placing the herrings in a nine-deep queue

right round the world might be impressive if we once began to realize what such a queue means. Let us see where the comparison came from. We may allow a herring and a half to the foot:

$$= 1\tfrac{1}{2} \times 5280 \text{ to the mile}$$
$$= 7920, \text{ or say } 8000 \text{ to the mile.}$$

Nine times round the world is $25,000 \times 9$ miles. So altogether there are $25,000 \times 9 \times 8000$ herrings,

$$= 1,800,000,000 \text{ herrings.}$$

That is a herring each for every man, woman and child in the world. It is astonishing how often such comparisons come unbidden, when one is used to considering things arithmetically.

To anyone who has considered what a million is, the number itself—not far short of 2000 millions—is sufficiently impressive. The mere counting of such a catch—if one did trouble to count them—would be a job for eternity rather than time. At the excessive rate, for herrings, of 100 per minute we found that it would take a month to count a million. The herring catch would take 1800 months or 150 years for one man to count it.

Let us try the effect of division on the number. If we omit very small children England has a population of about 36 millions—that number is convenient for division into 1800 millions:
$$1800 \div 36 = 50.$$

That is 50 herrings each. So that the population of England might have a herring each on almost every Friday of the year from the Yarmouth catch.

With the necessary information we could find how many boxes the herrings would fill, and how many railway waggons would be needed to carry them. The author of the first comparison started with the total catch—1800 millions. He imagined them head to tail so as to make the most of them.

1½ herrings to the foot = 8000 (about) to the mile. 1800 millions would stretch

$$\frac{1,800,000,000}{8000} \text{ miles}$$

$$= 225,000 \text{ miles,}$$

and there is then a ready comparison with the circumference of the world. There we have the comparison looked at first from one end and then from the other end.

A friend brought me the statement from a newspaper that during a certain snowstorm 2,000,000 tons of snow had fallen on Hull. He imagined that several noughts had been inadvertently added to the number, but he had apparently no means of checking it. After ten seconds' consideration I was able to assure him that there is nothing inherently improbable in the statement, and I also informed him that the area of Hull is round about 10,000 acres. My friend seemed to think that I was an arithmetical magician, or alternatively that I had worked out the problem previously.

All I had done was to divide by 2, and knock two noughts off 1,000,000 and thus obtain 10,000. So easily might a reputation be built up!

It was quite evident that there had been a very heavy fall of snow; otherwise it would not be worth while commenting on it. Let us see what we can do with the facts that most people know:

1 acre = 4840 square yards,
1 cubic foot of water weighs about 62½ lb.

We begin with an acre:

$$1 \text{ acre} = 4840 \text{ square yards}$$
$$= 4840 \times 9 \text{ square feet.}$$

An inch of rain over an acre would occupy a volume of:

$$4840 \times 9 \times \tfrac{1}{12} \text{ cubic feet.}$$

This volume of water would weigh:

$$4840 \times 9 \times \tfrac{1}{12} \times 62\tfrac{1}{2} \text{ lb.} = \frac{4840 \times 9 \times \tfrac{1}{12} \times 62\tfrac{1}{2}}{2240} \text{ tons}$$

$$= \frac{4840 \times 9 \times 125}{2240 \times 12 \times 2} \text{ tons}$$

$$= \frac{363,000}{3584} \text{ tons.}$$

For the purpose of a rough comparison that is near enough to 100 tons. So that an inch of rain on an acre of ground weighs 100 tons.

As every schoolboy learns 10 inches of snow is equivalent to an inch of rain. A fall of 10 inches of snow would therefore be equivalent to an inch of rain, and would therefore weigh 100 tons per acre. The encyclopaedia gives the area of Hull as 9042 acres. So that a fall of 20 inches of snow—a fall worth comment—would give a weight of nearly 2 million tons.

The process of thought—a matter of seconds—was: 20 inches of snow, a reasonably heavy fall, and easy to calculate with; equal to 2 inches of rain; weighs 200 tons per acre; divide 2,000,000 by 2 and knock off two noughts; gives 10,000—the approximate area of Hull. So much for magic!

At least one person in every newspaper office knows the following facts:

1 inch of rain weighs 100 tons per acre,

10 inches of snow = 1 inch of rain.

Whenever there is an exceptionally heavy fall of rain or snow
it is only necessary to look up the acreage in the encyclopaedia,
add two noughts for rain (or one nought for snow) and then
multiply by the number of inches of rain or snow. "Hull's
Two Million Tons of Snow." What a headline, and how
easy to calculate! It is no wonder the comparison is popular.
London is so big that the comparison is unusually striking.
The area of Greater London is about 450,000 acres. So that
an inch of rain on London weighs 45 million tons, and we
easily arrive at "London's Fifty Million Tons of Rain."

All the familiar comparisons are worked out in the way
we have just considered. They are matters of the simplest
arithmetic. Ingenuity comes in in choosing the most apt
comparison.

Light travels 186,000 miles per second. That is an almost
inconceivable speed. Seven and a half times round the
world in a second. That is a much more reasonable comparison than the same sort of comparison applied to a catch
of herrings. It is more reasonable because light does travel
great distances, whereas herrings do not form head to tail
queues round the world. It is reasonable also because a
second is the smallest unit of time with which most of us are
familiar, and the circumference of the earth is the greatest
length of which most people have any conception at all. It
is a small objection that the circumference of the earth is a
curve, whereas light travels in straight lines.

Time comparisons are often very effective because most
people have at least some sense of the passage of time. The
hours of work compel our attention to time, even if nothing
else does.

We often compare the speeds of light and sound. There is
an impressive comparison between the single second that

light takes to leap across 186,000 miles and the time that sound takes to travel the same distance. Sound travels a mile in 5 seconds. That is:

$$186,000 \text{ miles in } 186,000 \times 5 \text{ seconds}$$
$$= 930,000 \text{ seconds}$$
$$= 15,500 \text{ minutes}$$
$$= \text{about } 260 \text{ hours}$$
$$= \text{nearly } 11 \text{ days.}$$

So that sound takes 11 days to traverse the distance that light leaps across in a second. We are assuming of course the possibility of sound travelling such a distance; the fact that it is, to say the least, highly improbable that sound should travel such a distance, does not affect the validity of the comparison. We are comparing two speeds, and the comparison is:

$$\frac{\text{speed of light}}{\text{speed of sound}} = \frac{11 \text{ days}}{1 \text{ second}}$$

All we are saying is that the speed of light is as many times as great as the speed of sound, as 11 days is as great as 1 second.

The astronomical unit of the light-year employs the same sort of comparison. We get some sort of an idea of the distance that light travels in a second—$7\frac{1}{2}$ times the distance round the world—and then we have the comparison between a second and a year; we all have some feeling about that comparison, some sense of what it means.

Money is another means of getting effective and striking comparisons. Money comes so intimately into our lives that we all have some feeling about the value of money, even about such large sums as the national revenue. There is an

obvious comparison for distances, because we can find the
fares for the distances we want to compare.

Let us compare the fares: round the world, to the moon,
to the sun, and to the nearest star. We had better start with
very cheap fares, or the bigger ones may be impossibly big.
Let us say a thousand miles for a penny.

At this rate we can go round the world for 25 pence, or
say 2 shillings.

To the moon the fare is 238 pence, or say 1 pound.

To the sun the fare is 93,000 pence

$$= £387. \ 10s.$$
$$= \text{nearly } £390.$$

The distance of the nearest star is about 26 billion miles.
The fare is 26,000,000,000 pence. For the present purpose
we can take this as the convenient amount of 24,000,000,000
pence:
$$= £100,000,000.$$

And so we get the comparison:

Round the world: a florin.

To the moon: a pound.

To the sun: nearly £390.

To the nearest star: over a hundred million pounds.

If the population of the British Isles (say 45 millions) were
arranged in a line—we might allow a distance of half a yard
for each person—the line would extend:

$$22,500,000 \text{ yards} = 12,780 \text{ miles},$$

that is just about half-way round the world. At 50 miles an
hour a train would take 256 hours to travel this distance:

$$= \text{about } 10\tfrac{1}{2} \text{ days.}$$

The fare at a penny a mile would be about £53. At a cheap rate it would cost £53 to travel past the population arranged in a line.

If we arrange the population in a square, we might allow a square yard each. This would give a total area of 45,000,000 square yards:

$$= \frac{45,000,000}{1760^2} \text{ square miles}$$

$$= \text{about } 14\tfrac{1}{2} \text{ square miles.}$$

That is they could all be comfortably accommodated in a square with sides a little less than four miles long.

Allowing a cubic yard each we might find what fraction of a cubic mile the whole population would occupy:

$$\frac{45,000,000}{1760^3} = \text{about } \frac{1}{120}.$$

So that the whole population of the British Isles could be accommodated (uncomfortably) in the 120th of a cubic mile—or say a space one mile square and the 120th of a mile high:

$$1760 \div 120 \text{ yd} = \text{say } 14\tfrac{1}{2} \text{ yards.}$$

The whole population of the world (say two thousand millions, though the estimate is rather more) would occupy:

$$\frac{2,000,000,000}{1760^3} \text{ cubic mile}$$

$$= \cdot 367 \text{ cubic mile}$$

$$= \text{about } \tfrac{1}{3} \text{ cubic mile.}$$

Many of the comparisons we use are scale models of larger things. There is a famous model of the solar system in which the sun is represented by a sphere 2 feet in diameter. It is a matter of the simplest arithmetic to reduce the sizes and distances of the planets to this scale. The actual measurements may be found in an encyclopaedia or in a book on astronomy. They are:

	Distance (millions of miles)	Diameter (miles)
Sun	—	866,000
Mercury	36	2,800
Venus	67	7,500
Earth	93	8,000
Mars	142	4,200
Jupiter	483	88,000
Saturn	886	76,000
Uranus	1783	31,000
Neptune	2794	33,000

On the scale we are considering:

866,000 miles is represented by 24 inches,

$$8000 \text{ miles by } \frac{24 \times 8000}{866,000} \text{ inches}$$

$$= \cdot 22 \text{ in. (about } \tfrac{1}{4} \text{ in.).}$$

93,000,000 miles is represented by:

$$\frac{2 \times 93,000,000}{866,000} \text{ feet}$$

$$= \text{about } 210 \text{ feet}$$

$$= 70 \text{ yards.}$$

In the scale model therefore the earth is represented by a small sphere less than a quarter of an inch in diameter, at a distance of 70 yards from the 2 foot globe which represents the sun.

As soon as the earth has been reduced to scale the other planets are much easier. Consider the diameters first of all.

Mercury is a little more than a third of the earth, about ·08 or nearly a twelfth of an inch. Venus is a little less than the earth, say a fifth of an inch. Mars is about half the earth—an eighth of an inch. Jupiter is 11 times the earth—about 2½ inches. Saturn is 9½ times the earth—about 2 inches. Uranus is nearly 4 times the earth—nearly 9 tenths of an inch, and Neptune is a little more.

Now consider the distances. Mercury is at $\frac{36}{93}$ of the earth's distance; in the model it is at a distance of:

$$\frac{36}{93} \times 70 \text{ yards} = \text{about 27 yards.}$$

Venus $\frac{67}{93} \times 70$ yards = about 50 yards.

Mars $\frac{142}{93} \times 70$ yards = about 107 yards.

Jupiter $\frac{483}{93} \times 70$ yards = about 360 yards.

Saturn $\frac{886}{93} \times 70$ yards = about 667 yards.

Uranus $\frac{1783}{93} \times 70$ yards = about 1340 yards (say ¾ mile).

Neptune $\frac{2794}{93} \times 70$ yards = about 2100 yards (say 1¼ miles).

If one considers this model it can give an impressive idea of the emptiness of space. In a width of 2½ miles there are the two foot globe representing the sun, eight small globes ranging in size from the size of a large seed to the size of a tennis ball, and a few smaller bodies. The rest is emptiness.

We have to imagine this model of the solar system isolated in space. Suppose we try to fit the nearest star into the same model. Its distance is 26 billion miles (26×10^{12}).

866,000 miles is represented by 2 feet

$$26 \times 10^{12} \text{ miles by } \frac{2 \times 26 \times 10^{12}}{866,000} \text{ feet}$$

$$= \frac{2 \times 26 \times 10^{9}}{866 \times 5280} \text{ miles}$$

$$= 11,370 \text{ miles.}$$

That is to say we have the solar system represented by eight small spheres circling round a two foot globe—the outermost planet being represented by a one inch ball a mile and a quarter from the centre. If the model were fixed at the North Pole the nearest star would be as far away as the South Pole and nearly half as far again. We have to imagine a spherical space with a diameter nearly three times as great as that of the earth. At the centre of this immense space is the two foot sphere and the eight small bodies circling round it. Then all is emptiness till one spot on the circumference of the sphere is reached.

According to a report in *The Times*, Sir James Jeans stated in a lecture on "The Depths of Space":

"If we constructed a model on the scale of 1,000,000 light-years to the inch all the stars we could see with our unaided eyes were contained in a sphere less than one-hundredth of an inch in diameter—a mere speck of dust. But the telescopic universe occupied a sphere 80 ft. in diameter. In the model our whole galaxy was a small disc the size of a pinhead; our sun was a single electron, and the earth was one-millionth part of an electron."

Let us examine this comparison. The diameter of the sphere of visible stars is given as 6000 light-years:

1,000,000 light-years is represented by 1 inch,

6000 light-years is represented by $\frac{6}{1000}$ inch,

$$= \frac{1}{166 \cdot 7} \text{ inch.}$$

So that a two-hundredth of an inch would be a better rough statement of the diameter.

80 feet would represent 80 × 12 million light-years, or about a thousand million light-years. This seems to be five

times the accepted range of two hundred million light-years for the diameter of the telescopic stars.

The galaxy may be the lesser Milky Way or the greater Milky Way. The former has a diameter of 20,000 light-years; this would be represented in the scale by one-fiftieth of an inch. The greater Milky Way has a diameter ten times as great, so that it would be represented by one-fifth of an inch. I have measured a pinhead and find it to be one-fifteenth of an inch across, so that neither a fifth nor a fiftieth of an inch appears to be a probable diameter for a pinhead.

The diameter of the sun is about 866,000 miles, say roughly one million miles; this is near enough for our present purpose. A light-year is about 6 billion miles.

One inch on the scale = 1 million light-years
= 6 trillion miles.

Hence 1,000,000 miles (diameter of sun roughly) is represented by $1 \div 6$ billion inch:

$$= \cdot 00000000000\tfrac{1}{6} \text{ inch}$$
$$= \cdot 00000000000\tfrac{1}{6} \times 2 \cdot 54 \text{ cm.}$$
$$= \cdot 0000000000004 \text{ cm.,}$$

and this is the diameter of an electron to a close enough approximation.

We now, rather unaccountably, after dealing with lengths, switch over to volumes. The diameter of the earth is very roughly 1/100 that of the sun, so that the volume is (three times more roughly) 1/1,000,000 that of the sun. That is the volume of the earth is represented by the millionth part of an electron.

The comparison of the sun with an electron comes in so aptly that it seems as if the model began there. I suggest that it would be better to begin with the earth as an electron.

We have a vague feeling of the size of the earth, and some conception of the smallness of an electron; but the millionth part of an electron seems an unnecessary burden. Also, we can keep to diameters, without the complication of mixing diameters and volumes. We need only multiply the scale by 100. We then have:

> Scale: 1 inch = 10,000 light-years.
> Earth: an electron.
> Naked-eye star sphere: $\frac{3}{8}$ inch diameter.
> Galaxy (greater Milky Way): 20 inches diameter.
> Diameter of telescopic range: 20,000 inches
> $\qquad\qquad$ = about $\frac{1}{3}$ mile.

That seems to me a more understandable comparison.

We have been scaling down large quantities in order to get familiar comparisons within the range of our understanding. When we come to very small numbers it is necessary to exaggerate them, to multiply them, in order to get comparisons we can appreciate.

A molecule of hydrogen is a very small thing. Its diameter is about $2\frac{1}{2} \times 10^{-8}$ centimetre or 10^{-8} inch = ·00000001 inch. A hundred millions of them, side by side, would stretch an inch. Suppose then we multiply by a hundred millions, the hydrogen molecules become little spheres an inch in diameter.

What does 10^8 suggest? It is a big number. We want something containing a hundred million of some unit of length. Suppose we try the diameter of the earth:

> $= 8000$ miles
> $= 8000 \times 5280$ feet
> $= 42,240,000$ feet
> $=$ about $500,000,000$ inches
> $=$ about 5×10^8 inches.

Thus, if we magnify a sphere 5 inches in diameter a hundred million diameters we shall have a sphere as big as the world. And we have seen that this magnification would change the hydrogen molecules into spheres 1 inch in diameter.

We now have the comparison: if a sphere of hydrogen 5 inches in diameter were magnified to the size of the earth, the molecules would be the size of 1 inch spheres.

The number of molecules in a gram of hydrogen is so immense that it is almost impossible to get a comparison. The number is:

$$3 \times 10^{23}$$

The width of a molecule of hydrogen is very small, and it might be thought that by placing them side by side in a row we should find that the molecules in a gram of hydrogen stretch an appreciable but not a considerable distance. But see what happens.

The width of a molecule of hydrogen is 10^{-8} inch. So the total length of the molecules in a gram is:

$$10^{-8} \times 3 \times 10^{23} \text{ inches,}$$

$= 3 \times 10^{15}$ inches

$= 2\frac{1}{2} \times 10^{14}$ feet

$=$ about 5×10^{10} miles

$= 50,000,000,000$ miles

$=$ nearly ten times the width of the whole solar system.

And after all we should expect some such result. We have already seen how far things are apt to extend when they are placed side by side or end to end. Imagine the molecules packed evenly in a cubic inch. In one row along one edge we get 10^8 (a hundred million) molecules, stretching a length of 1 inch. But then in one face of the cube (in a single layer) there are 10^8 of these rows; and there are 10^8 layers. So actually we have to multiply an inch twice by 10^8, that is by 10^{16} (ten thousand billions); so the length is ten thousand billion inches. Actually the molecules in hydrogen are not so tightly packed at standard temperature and pressure; nevertheless the total width of all the molecules is, as we have seen, enormous.

On the whole we had better return to our former comparison and say that the number of molecules in a gram of hydrogen (a 9 inch cube of hydrogen) is not far distant from the number of 1 inch cubes that would fill the whole volume of the earth—a sufficiently impressive picture of a very big number.

I have purposely selected extremely large and extremely small quantities to explain how the familiar comparisons are arrived at. It is comparatively easy to find comparisons for distances less large than stellar distances, or less small than molecules. Many large numbers we can compare with the diameter or circumference of the earth, the width of the Atlantic, the length of England, and so on. We can place large numbers of objects end to end, to make them stretch an impressively great distance, or we can pack them in a cube or sphere so as to get them into a compact space. Small objects we can magnify till they assume appreciable size. But however we manipulate the quantities, the whole thing is a matter of proportion—multiplication and division.

It is a common, and a very useful, device to illustrate statistics by means of diagrams. Such diagrams often show at a glance what is contained in a mass of numbers; the pictorial comparison of numbers can be very striking. If the numbers are represented by lines, or by oblongs of equal width, or by units of equal size, there can be no objection to them. But as soon as diagrams begin to acquire more than one dimension they can be very misleading.

The big loaf is twice the height of the little one, and it might be speciously argued that the diagram is meant to represent one quantity of bread twice as much as another quantity. But the big loaf is also twice as long and twice as wide as the little one, so that actually it represents a loaf, not twice, but eight times as big as the other.

Drawings of men, ships, trains, bales of goods, etc. are often used to illustrate statistics. Railway trains can be used as if they were lines, because only the length of the train need vary. Bales of goods can be used in the same way, and so can other things where each separate object is treated as a unit.

When the objects are magnified or diminished to represent larger or smaller quantities we can no longer treat them as if they were lines. A man drawn 4 inches high has much the same shape as a man drawn 1 inch high; his volume is not 4 times as great, but $4 \times 4 \times 4 = 64$ times as great.

If numbers are to be represented by lines the lines should be drawn proportional to the numbers they represent. The following diagram would represent lengths of 4 units and

3 units, or any quantities in the proportion $4 : 3$. But these

drawings would represent, not $4 : 3$, but $4^3 : 3^3$, that is $64 : 27$.
$\frac{4}{3} = 1\frac{1}{3}$; $\frac{64}{27} = 2\frac{10}{27}$ (about $2\frac{1}{3}$); so that there
is a considerable difference.

We can use drawings of obviously
flat surfaces—squares, oblongs, triangles,
circles. In such drawings it is the areas
that represent the numbers or quantities
and therefore the areas should be proportional to the quantities. We know
that areas are proportional to the squares of similar lines
(the length of a side of a square, the radius of a circle, etc.).

$$\text{Area} = c \text{ length}^2,$$

and as the area is to be proportional to the quantity to be
represented:
$$\text{quantity} = c \text{ length}^2,$$
or
$$\sqrt{\text{quantity}} = \sqrt{c} \text{ length}$$
$$= k \text{ length}.$$

Thus in diagrams in which numbers are represented by
areas the lengths of the lines should be proportional to the
square roots of the quantities. Suppose we wish to represent
64 and 81 by means of circles. The radii of the
circles should be in the proportion $\sqrt{64} : \sqrt{81} = 8 : 9$.

When we use drawings of the same solid (e.g.
the big and little loaf) to represent numbers or
quantities the volumes should be in proportion
to the numbers or quantities represented. And
the volumes as we know are proportional to the cubes of
similar lengths:
$$\text{volume} = c \text{ length}^3.$$
As before:
$$\text{quantity} = c \text{ length}^3,$$
or
$$\sqrt[3]{\text{quantity}} = k \text{ length}.$$

That is similar lines in the diagrams should be proportional to the cube roots of the quantities.

Suppose we wish to represent 125 and 64 by means of cubes (or spheres, or other similar solids), the lines in the solids (e.g. edges of the cubes) should be in the proportion $\sqrt[3]{125} : \sqrt[3]{64} = 5 : 4$.

Suppose we wish to represent 64 and 729 pictorially. We can choose between lines, surfaces, and volumes. Lines may be anything that varies only in length; they should be proportional to 64 and 729, that is the second line should be nearly $11\frac{1}{2}$ times as long as the first. Areas may have no apparent thickness, or the thickness may be the same for both; lines in the areas are in the proportion $\sqrt{64} : \sqrt{729} = 8 : 27$, that is lines in the second drawing should be nearly three and a half times as long as similar lines in the first. When volumes are used (e.g. men, ships, or trains—magnified or diminished) the lengths of similar lines should be in the proportion $\sqrt[3]{64} : \sqrt[3]{729} = 4 : 9$; that is lines in the second drawing should be $2\frac{1}{4}$ times the length of similar lines in the first drawing. The proportions are:

for lines: $64 : 729 = 1 : 11\cdot4$,
for areas: $8 : 27 = 1 : 3\cdot4$,
for volumes: $4 : 9 = 1 : 2\cdot25$.

Before accepting a pictorial representation of numbers it is well to see that it is correct.

It is very commonly thought that if lines are reduced to one-tenth everything else is reduced in the same proportion. We have seen that this is a fallacy. Areas are reduced, not to a tenth, but to a tenth squared, that is to a hundredth $(\frac{1}{10} \times \frac{1}{10} = \frac{1}{100})$. Volumes are reduced to the cube of a tenth, that is to a thousandth. The weight of an object depends partly on its volume and partly on the kind of material of which it is composed (that is, on its density); if the material is the same in both cases, then the weight also is reduced to one-thousandth.

In *Gulliver's Travels* Swift describes diminutive men. But his Lilliputians are not really diminutive men; they are ordinary men looked at through a reducing glass. As an example of the difference—under the reducing glass all lengths are reduced in proportion, and speeds are correspondingly reduced. But speeds are not so reduced. An elephant may be twenty times as high as a rabbit, but it does not therefore run twenty times as quickly; the strength of an elephant is considerably greater than that of a rabbit, but so also is the mass which the strength is required to move. A diminutive man (under the reducing glass) might be killed by falling from a height of a foot; but we know that a mouse can fall half a mile without being injured. As another small point of difference, the amount of food required by a small creature is far greater in proportion to its size than that required by a larger creature.

The men in Brobdingnag are ordinary men magnified. In *The Food of the Gods* Mr Wells also describes ordinary men and other creatures, as seen through a magnifying glass. The weight of a sixty foot man would actually be not ten times, but the cube of ten, that is a thousand times as great as that of a six foot man. He would weigh something like seventy-five tons. The soles of his feet would be only a hundred ($= 10^2$) times as great as those of a six foot man; so that each square inch of surface would have to support a pressure ten times as great as in the case of an ordinary man. The materials of the human body could not support such a pressure.

Sixty foot men would be clumsy, coarse, lumbering creatures, with hides thicker than those of elephants; they could not walk upright. The magnified dragonfly larvae of *The Food of the Gods* have their strength and their biting power multiplied without apparently any change in the materials of which their bodies are composed.

In *Travel Tales of Mr Joseph Jorkens* Lord Dunsany introduces us to the planet Eros, with a herd of elephants small enough to go into match-boxes. One of these elephants is sufficiently powerful to break out of a match-box. Once again these are ordinary elephants seen through a reducing glass; indeed Lord Dunsany says as much when he compares the elephant in a match-box to a real elephant in a wooden hut. As we have seen, such a comparison is fallacious. We should make an allowance also for the small weight of any creature on so small a planet. So far as we can see the bigger a planet is the smaller would be the inhabitants because of the increase in weight. So far as I know no one has considered the effect of weight on the strength of muscles. Mr Wells gives his men in the moon strength undiminished by decrease in weight; Lord Dunsany gives his diminutive elephants a strength which also is not diminished by decrease in weight.

Truly the ways of creators are hard. There are so many things one is apt to leave out of account. It is so hard to get *all* one's proportions right; that is why one has to be dubious about untested utopias.

The "fallacy of the model" is a thing one has always to be on guard against. We are very apt to forget that when lengths are doubled areas are increased fourfold and volumes eightfold; if the same materials are used then the weight also is increased eightfold. The lifting power of an aeroplane depends on the spread of the wings, that is on the square of lengths. If we double the lengths in an aeroplane the area (and with it the lifting power) is four times as great; but the volume and the weight to be lifted are eight times as great. And so it arises that it is comparatively easy to fly a small aeroplane; but every increase in size increases the difficulty of flight. Model aeroplanes were made and flown long before there was an engine powerful enough to drive a large aeroplane.

Attempts to imitate the flight of birds were based on the fallacy of the model. Man was to become a magnified bird. But a magnified bird could not fly—if the magnification were sufficiently great. As we have seen doubling the length of a bird would increase the wing-spread fourfold, but it would make the weight to be lifted eight times as great.

VI

Proportion in Triangles

THE whole subject of trigonometry is based almost entirely on the study of triangles, and one is apt to exclaim "Why all this fuss about triangles? Squares or oblongs would be much simpler to deal with." Well, there are several very good reasons why the study of triangles is so important. A triangle is the simplest straight-lined figure. It is not deformable. If the corners are pinned together the shape of the triangle is fixed; this is not true of a square or an oblong. Carpenters

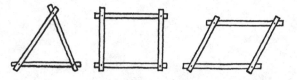

and engineers use this fact to strengthen their constructions. A piece of wood or metal fixed across a corner so as to make a triangle makes the angle of the corner unchangeable without breaking the materials. And it is easy to measure a triangle. When we have measured the base and the two angles at the base, we know the triangle completely; we can calculate the remaining angle, the lengths of the other sides, and the area. And we can make a triangle by joining any three points not in a straight line; so that when we measure by means of triangles we can use any convenient points.

If you examine these two triangles you will see that *DEF* is a magnified version of *ABC*. The two triangles have the

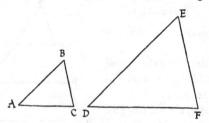

same shape. Angle *A* = angle *D*; angle *B* = angle *E*; angle *C* = angle *F*. We say the triangles are similar.

When we magnify we magnify all the lines alike. If $DE = 2AB$, then $DF = 2AC$, and $EF = 2BC$. If $DE = 1\frac{1}{2}AB$, then $DF = 1\frac{1}{2}AC$, and $EF = 1\frac{1}{2}BC$. We can say in general that in any two similar triangles corresponding sides are in proportion:

$$DE = k \times AB; \quad DF = k \times AC; \quad EF = k \times BC$$

(*k* is a constant: 2, or $1\frac{1}{2}$, or whatever it may be).

All similar lines in the two triangles are in the same proportion; this is no more than to say that the whole triangle is magnified the same amount. *AL* and *DP* are perpendiculars from *A* and *D* to the opposite sides:

$$DP = k \times AL,$$

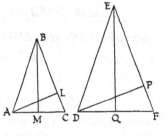

BM and *EQ* are lines from *B* and *E* to the mid-points of the opposite sides:

$$EQ = k \times BM,$$

and so on with all other similar lines.

In the triangle ABC we have DE and FG drawn parallel to BC. It should be clear that

$$\angle B = \angle AFG = \angle ADE.$$

The angles along AC are also equal.

We have therefore three similar triangles ADE, AFG and ABC —one over another.

If $AD = \frac{1}{3}AB$, then $AE = \frac{1}{3}AC$, and $DE = \frac{1}{3}BC$.

If $AF = \frac{2}{3}AB$, then $AG = \frac{2}{3}AC$, and $FG = \frac{2}{3}BC$.

The areas of the three similar triangles are proportional to the squares of similar sides:

$$\text{area of } ADE = k \times DE^2,$$
$$\text{area of } AFG = k \times FG^2,$$
$$\text{area of } ABC = k \times BC^2.$$

We found a moment ago that $DE = \frac{1}{3}BC$ and $FG = \frac{2}{3}BC$. Hence:

$$\text{area of } ADE = k \left(\tfrac{1}{3}BC\right)^2 = k \times \tfrac{1}{9}BC^2,$$

and $\quad\text{area of } AFG = k \left(\tfrac{2}{3}BC\right)^2 = k \times \tfrac{4}{9}BC^2.$

Hence area ADE : area AFG : area ABC

$$= k \times \tfrac{1}{9}BC^2 : k \times \tfrac{4}{9}BC^2 : k \times BC^2$$
$$= \tfrac{1}{9} : \tfrac{4}{9} : 1$$
$$= 1 : 4 : 9 \text{ (multiplying by 9).}$$

The proportion of the lines is $1 : 2 : 3$.

The proportion of the areas is $1 : 4 : 9 = 1^2 : 2^2 : 3^2$.

The idea of similar triangles may be applied in many different ways.

Look at the two chords AB and CD which cross at O. We have joined AC and BD. Now look at the arc $ACBD$.

There are two angles, C and B, which have their apices on the arc and the ends of their arms at A and D (the ends of the arc). The angles C and B are therefore equal. The angles marked Y at O are also of course equal.

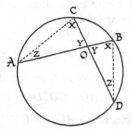

Now look at the two triangles AOC and BOD. These two triangles are similar, because the sets of angles marked X, Y, and Z are equal in pairs. Let us pick out the corresponding lines in these two similar triangles.

AC and BD—opposite the equal angles Y,
OC and OB—opposite the equal angles Z,
OA and OD—opposite the equal angles X.

So $AC = k \times BD$; $OC = k \times OB$; $OA = k \times OD$.

Look at the two parts of AB; they are OA and OB:

$$OA.OB = k \times OD.\frac{OC}{k}$$
$$= OD.OC.$$

That is the well-known theorem that the rectangle contained by the parts of one of two intersecting chords is equal to the rectangle contained by the parts of the other chord.

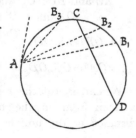

There is a very elegant extension of this theorem. Imagine one of the chords swung round on one of its ends as a pivot whilst the other chord remains fixed.

Finally the chord AB becomes the tangent at A. If we continue DC outward it cuts the tangent at O. A and B are now the same spot.

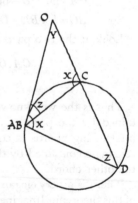

Now we had

$$AO . BO = CO . DO.$$

If we write A instead of B we have:

$$AO . AO = CO . DO \text{ (or } OC . OD).$$

Hence $AO^2 = OC . OD.$

The square on the tangent from O is equal to the rectangle contained by OC and OD, where OD is a line drawn across the circle from O.

The parallel between the two theorems is exact. If we join AC and BD as before, we shall have two similar triangles AOC and BOD, exactly as before. The equal angles at O become the single angle at O. The equal angles ACO and OBD are now the angles X.

The corresponding sides are:

$$AC \text{ and } BD; \ OC \text{ and } OB;$$
$$OA \text{ and } OD.$$

$$OA . OB = k \times OD . \frac{OC}{k} = OD . OC.$$

$$\therefore \ OA^2 = OC . OD.$$

We can use this theorem to find the distance of the horizon as seen from any height. Suppose we want to find the distance of the horizon as seen from a height OC. We imagine the line OC continued downwards through the centre of the earth to the antipodes. Then we have:

$$OA^2 = OC.OD.$$

OA is the distance we want to find. OC is the height from which we are finding the distance of the horizon. OD is the diameter of the earth ($OD = 8000$ miles; we neglect the small distance OC because this is very small compared with CD; it is usually less than the error made in taking the diameter of the earth as 8000 miles).

Hence $OA = \sqrt{OC.8000}$ miles. We must of course have OC in miles.

Let us find the distance of the horizon to a man standing at a height of 100 feet ($= \frac{100}{5280}$ miles) above the sea.

$$OA = \sqrt{\frac{100 \times 8000}{5280}} = \sqrt{\frac{80000}{528}}$$

Here is the working:

```
        151·52                151·52 (12·3
   528) 80000                 144
        528             243  | 752
        2720                 | 729
        2640                   ==

         800
         528

         272
         ==
```

The distance of the horizon is a little more than 12·3 miles.

From the top of a mountain 10,000 feet high (say 2 miles high) the greatest distance at which it is possible to see anything at sea-level is:

$$\sqrt{2 \times 8000} \text{ miles}$$
$$= 40 \sqrt{10} \text{ miles}$$
$$= 40 \times 3 \cdot 16 \text{ miles}$$
$$= \text{about } 126\tfrac{1}{2} \text{ miles.}$$

Suppose we want to find the greatest distance from which the top of one mountain can be seen from the top of another. The diagram shows that we have to add the distances of the two horizons $OQ + QP$. Thus the top of a mountain 1 mile high could just be seen low down on the horizon from the top of a mountain 2 miles high, if the distance between the two mountains is:

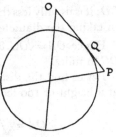

$$\sqrt{2 \times 8000} + \sqrt{1 \times 8000} \text{ miles}$$
$$= \text{about } 126\tfrac{1}{2} \text{ miles} + 89\tfrac{1}{2} \text{ miles}$$
$$= \text{about } 216 \text{ miles.}$$

We are assuming of course that there is an uninterrupted view, and ignoring effects of refraction of light.

The greatest distance from which it would be possible to see the top of a mountain 5 miles high from the top of a similar mountain is:

$$2 \sqrt{5 \times 8000} \text{ miles}$$
$$= 400 \text{ miles.}$$

Let us look now at another famous theorem which is readily deduced by considering similar triangles.

ABC is a right-angled triangle; the right angle of course is at *A*. *AD* is the perpendicular from *A* to *BC*.

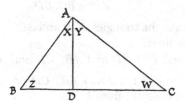

There are three triangles in the figure—*ABC*, *ABD*, *ACD*. They are all three similar. We know that the three angles of a triangle add up to two right angles. Hence the angles marked Z and W add up to one right angle (angle *BAC* is the other, right, angle of the triangle *ABC*).

$$Z + W = 1 \text{ rt. angle}$$
$$Z + X = 1 \text{ rt. angle}$$
$$\therefore \quad W = X.$$
$$W + Y = 1 \text{ rt. angle}$$
$$\therefore \quad Y = Z.$$

Look at the similar triangles *ABD* and *ABC*. Pick out the pairs of similar lines:

> *AB* and *BC*—opposite the right angles,
> *AD* and *AC*—opposite angles *Y*,
> *BD* and *AB*—opposite angles *X*.

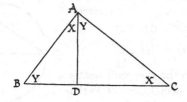

Hence: $AB = k \times BC$; $AD = k \times AC$; $BD = k \times AB$,

$$\therefore \quad k \times AB . AB = k \times BC . BD$$
$$\therefore \quad AB^2 = BC . BD.$$

Now let us take the triangles ACD and ABC. We have the pairs of similar lines:

$$AC \text{ and } BC; \quad AD \text{ and } AB; \quad CD \text{ and } AC.$$

Hence $AC = c \times BC$; $AD = c \times AB$; $CD = c \times AC$
(c is a different constant from k);

$$\therefore \quad c \times AC . AC = c \times BC . CD,$$
$$\therefore \quad AC^2 = BC . CD.$$

If we add the two results we have:

$$AB^2 + AC^2 = BC . BD + BC . CD$$
$$= BC \, (BD + CD)$$
$$= BC^2.$$

This is the famous theorem of Pythagoras.

There is another pair of similar triangles we have not yet considered. In triangles ABD, ACD we have the pairs of similar lines AB and AC, AD and DC, BD and AD.

Hence $AB = l \times AC$; $AD = l \times DC$; $BD = l \times AD$
(l is another constant);

$$l \times AD . AD = BD . l \times DC,$$
$$\therefore \quad AD^2 = BD . DC.$$

We have already got this result because we can draw a circle on BC as diameter, and this circle passes through A. We have:

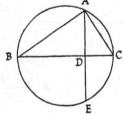

$$AD . DE = BD . DC,$$
$$\therefore \quad AD^2 = BD . DC,$$

which links up the theorems neatly.

There is a special case of Pythagoras' theorem which has been known

from very ancient times, and which probably suggested the
general theorem. This is the case where
the hypotenuse is 5 units long and the
shorter sides 4 and 3:

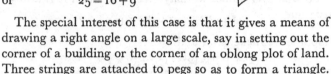

$$5^2 = 4^2 + 3^2$$
or $\qquad 25 = 16 + 9$

The special interest of this case is that it gives a means of
drawing a right angle on a large scale, say in setting out the
corner of a building or the corner of an oblong plot of land.
Three strings are attached to pegs so as to form a triangle.
The exact lengths between the pegs
are in the proportions 5 : 4 : 3. Two
of the pegs (say A and B—on one of
the shorter sides) are pressed into
the ground along one side of the
square or oblong; the string between
them is drawn out tight. The third

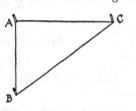

peg C is then drawn out till both strings are taut. The strings
may be 5 feet, 4 feet, and 3 feet, or $7\frac{1}{2}$ feet, 6 feet, and
$4\frac{1}{2}$ feet; any convenient multiples of 5, 4 and 3 will do, but
the greater the lengths (in reason) the less is the liability to
error.

It is rather amusing to hunt for sets of numbers with the
Pythagorean relation. Most people know the two sets:

$$5^2 = 4^2 + 3^2$$
and $\qquad 13^2 = 12^2 + 5^2$

The following method will enable anyone who wishes to do
so to find further sets (the equation gives them all):

$$(m^2 + n^2)^2 \equiv (m^2 - n^2)^2 + (2mn)^2$$

The three lines (in preference to the usual two lines for
"equals") draw attention to the fact that the two sides of

the equation are identically equal—two ways of writing the same thing. This can be shown by multiplying out:

$$(m^2 + n^2)^2 = m^4 + 2m^2n^2 + n^4,$$

and $(m^2 - n^2)^2 + (2mn)^2 = m^4 - 2m^2n^2 + n^4 + 4m^2n^2$
$$= m^4 + 2m^2n^2 + n^4.$$

So whatever values we may give to m and n it is always true that:
$$(m^2 + n^2)^2 = (m^2 - n^2)^2 + (2mn)^2$$

Suppose we make $m = 2$ and $n = 1$:

$$(2^2 + 1^2)^2 = (2^2 - 1^2)^2 + (2 \times 2 \times 1)^2$$

or $$5^2 = 3^2 + 4^2$$

If we make $m = 4$ and $n = 3$:

$$(4^2 + 3^2)^2 = (4^2 - 3^2)^2 + (2 \times 4 \times 3)^2$$

or $$25^2 = 7^2 + 24^2$$

There are two interesting facts that may be verified in every case:

1. The length of the hypotenuse is always the sum of two square numbers:

$$5 = 4 + 1 = 2^2 + 1^2$$
$$13 = 9 + 4 = 3^2 + 2^2$$
$$25 = 16 + 9 = 4^2 + 3^2$$

An odd fact, till we remember that we get the length of the hypotenuse from $m^2 + n^2$, that is the sum of two squares.

2. One side must be a multiple of 5, one must be a multiple of 4, and one must be a multiple of 3.

In $5^2 = 4^2 + 3^2$ that is clear enough.

In $13^2 = 12^2 + 5^2$, the 12 happens to be a multiple of both 4 and 3.

In $41^2 = 9^2 + 40^2$, the 40 happens to be a multiple of both 4 and 5. Notice also that $41 = 5^2 + 4^2$.

We will look now at a very interesting set of similar triangles. We have the three similar right-angled triangles

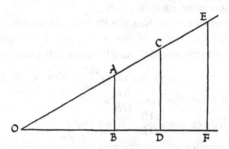

ABO, CDO, EFO. We can draw as many such triangles as we like by dropping perpendiculars from *OE* to *OF*.

There are three sets of similar lines:

the three perpendiculars: *AB, CD, EF,*
the three bases: *OB, OD, OF,*
the three hypotenuses: *OA, OC, OE.*

Since the triangles are similar:

$$AB = k \times CD = l \times EF,$$

and we can add

$$= m \times GH \text{ (where } GH \text{ is another perpendicular)}$$
and so on.

Also $OB = k \times OD = l \times OF = m \times OH.$

Now divide the quantities in the first set by those in the set below:

$$\frac{AB}{OB} = \frac{k \times CD}{k \times OD} = \frac{l \times EF}{l \times OF} = \frac{m \times GH}{m \times OH} = \text{etc.}$$

or $\qquad \dfrac{AB}{OB} = \dfrac{CD}{OD} = \dfrac{EF}{OF} = \dfrac{GH}{OH} = \text{etc.}$

That is the ratio $\dfrac{\text{perpendicular}}{\text{base}}$ is the same for every perpendicular and its own base.

The size of this ratio does not depend, as we have seen, on the lengths of the sides of the right-angled triangle. But it does depend on the size of the angle between the two long arms.

In this figure we have the ratio $\dfrac{AB}{OB}$. Now make the angle bigger; we have the ratio $\dfrac{KB}{OB}$ which is obviously bigger than $\dfrac{AB}{OB}$. If we make the angle smaller we get the ratio $\dfrac{LB}{OB}$, which is less than $\dfrac{AB}{OB}$.

The ratio $\dfrac{\text{perpendicular}}{\text{base}}$ is called the *tangent* of the angle at O. We have seen that for any angle the tangent is the same, no matter how long we may draw the arms of the angle. We have also seen that a bigger angle has a greater tangent than a smaller angle.

Every angle has its tangent, and every tangent has its angle. Which angle has 2·3 for its tangent? We could of course look it up in a book of tables. But we can also find it out by drawing the angle and measuring it with a protractor.

Look at this diagram. We draw AB as a base line, one unit long (1 inch, or 2 inches if we keep to 2 inches as a unit). We draw BC at right angles to AB and make it 2·3 units long (2·3 inches, or 4·6 inches if we keep to 2 inches as the unit). Then we join AC, and measure the angle CAB.

The tangent of CAB is $\dfrac{2\cdot3}{1}$ or $\dfrac{4\cdot6}{2}=2\cdot3$. What is the use of tangents? Why go to the trouble of finding them, when we can measure the angles themselves?

Well, in measuring land it is distances we want. The measurement of angles only comes in incidentally. It is nearly always easier to measure an angle than to measure a line. In measuring an angle we turn the telescope of a theodolite to one point and then swing it round to another point; the angle between the two points can thus be measured with great accuracy. Whereas to make a measurement of length across uneven ground, perhaps through a marsh or over land intersected by rivers, would be a matter of extreme difficulty.

We begin by making one measurement, as conveniently and as accurately as possible, on level ground. After that it is only necessary to measure angles.

Suppose we have measured the line AB, say 1000 feet. C is a convenient distant point, and we want to find the distance of C from A and B. We measure the angles at A and B. We fix the theodolite exactly over A, point it to B and then 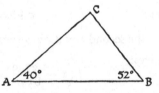 swing it round to point at C; this gives a measure of the angle BAC. A similar measurement is made of the angle at B.

If we like we can draw the triangle to scale. We draw AB to scale, construct the measured angles at A and B, and draw the lines to meet at C. Then we can measure the lengths of AC and BC from the scale.

It is a useful rule that if we have sufficient information to draw a figure to scale, this information is also sufficient to enable us to calculate lengths and angles in the figure.

CD is the perpendicular from *C* to *AB*. We have called the length of *CD h*. *a* and *b* are the lengths of *AD* and *BD*:

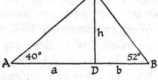

$$\tan A = \tan 40^\circ = \frac{h}{a}.$$

From a table of tangents we find that $\tan 40^\circ = \cdot 8391$,

$$\therefore \quad \cdot 8391 = \frac{h}{a} \text{ or } h = \cdot 8391a.$$

Tan $52^\circ = 1 \cdot 2799$, and so $h = 1 \cdot 2799b$.
    Thus we have:
$$h = \cdot 8391a = 1 \cdot 2799b.$$

We know also that $a + b = 1000$ feet. (This was the measured length of *AB*.)
    We now have two simple equations to find *a* and *b*:

$$\cdot 8391a = 1 \cdot 2799b,$$

and $$a + b = 1000 \text{ ft.}$$

In the second equation we write for *a* its value $\dfrac{1 \cdot 2799}{\cdot 8391} b$. This gives:

$$b \left(1 + \frac{1 \cdot 2799}{\cdot 8391} \right) = 1000 \text{ ft.}$$

$$b = \frac{1000}{1 + \dfrac{1 \cdot 2799}{\cdot 8391}} \text{ ft.}$$

$$= 396 \cdot 0 \text{ ft.}$$

$$a = 1000 - 396 \cdot 0 \text{ ft.}$$

$$= 604 \cdot 0 \text{ ft.}$$

$$h = \cdot 8391a = 506 \cdot 8 \text{ ft.}$$

Now $\qquad\qquad\qquad AC^2 = a^2 + h^2,$

and $\qquad\qquad\qquad AC = \sqrt{a^2 + h^2}$

$$= \sqrt{604 \cdot 0^2 + 506 \cdot 8^2}$$
$$= 788 \cdot 5 \text{ ft.}$$
$$BC^2 = b^2 + h^2,$$

and $\qquad\qquad\qquad BC = \sqrt{b^2 + h^2}$

$$= \sqrt{396 \cdot 0^2 + 506 \cdot 8^2}$$
$$= 643 \cdot 2 \text{ ft.}$$

As soon as we have found the lengths of AC and BC we can use these lengths as bases for finding the distances of places still farther away.

The problem has been worked out with the help of tangents only, in order to show the advantage of using other ratios.

$\dfrac{AB}{OB}$ is the tangent of angle O: tan O.

$\dfrac{AB}{OA} \left(\dfrac{\text{perpendicular}}{\text{hypotenuse}}\right)$ is called the sine: sin O.

$\dfrac{OB}{OA} \left(\dfrac{\text{base}}{\text{hypotenuse}}\right)$ is called the cosine: cos O.

(Sine—pronounced 'sine', written 'sin' before an angle; just as tangent is written 'tan' and cosine 'cos'.)

In the same way as for the tangent we can show that the sine and cosine are invariable for any angle. There are tables which give the sines and cosines of angles from $0°$ to $90°$. There is an interesting point about sines and cosines. The hypotenuse is the longest line in a right-angled triangle, and so neither sine $\left(\dfrac{\text{perpendicular}}{\text{hypotenuse}}\right)$ nor cosine $\left(\dfrac{\text{base}}{\text{hypotenuse}}\right)$ of any angle can be greater than 1.

Look again at the problem.
We used the tangents to find:

$$a = 604 \cdot 0 \text{ feet},$$
$$b = 396 \cdot 0 \text{ feet}.$$

We want to find AC.

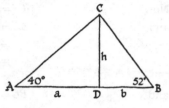

Now $\dfrac{AD}{AC} = \cos 40° = \cdot 7660$

(from the table of cosines).

Hence $$AC = \frac{a}{\cdot 7660} = \frac{604 \cdot 0}{\cdot 7660}$$

$$= 788 \cdot 5 \text{ feet}.$$

And $$\frac{BD}{BC} = \cos 52° = \cdot 6157.$$

Hence $$BC = \frac{b}{\cdot 6157}$$

$$= \frac{396 \cdot 0}{\cdot 6157} \text{ feet}$$

$$= 643 \cdot 2 \text{ feet}.$$

Many people find it extraordinary that it should be possible to measure the distances of sun, moon and stars. It cannot be possible, they say, to measure the distance of an object which we cannot reach. On the contrary, it is a familiar thing in surveying to measure the distance of a distant point without going near it. We have just shown how this can be done. The problem of measuring the distance of the heavenly bodies is essentially the same. We want a base line, and then we have to measure the angle between a telescope pointing at the object from one end of the base and a telescope pointing to it from the other end of the base. The chief trouble is to find a base long enough. A base of a thousand feet would be of little use, because the angle would be so small as to be immeasurable.

The most convenient base for measuring the distance of the sun is the diameter of the earth. Imagine a telescope fixed at the equator. It is pointed at a fixed point on the sun (say an edge of a sun spot). Twelve hours later the telescope will have moved right round with the rotating earth, so that it will be 8000 miles from its former position. The telescope will no longer point to the same spot on the sun; it has to be moved through a small angle to bring it back to the spot. The movement of the telescope will show what this angle is. Careful measurement shows that it is 17·62 seconds of arc. A second is the 1/3600 part of a degree, so that the measurement requires very careful work.

The rest is a matter of simple arithmetic:

BC is the diameter of the earth. BAC is the measured angle, and we want to find the length AB (the distance from earth to sun).

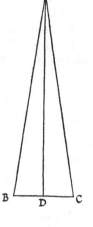

We have: $\sin BAD = \dfrac{BD}{AB}$

$$= \frac{4000 \text{ miles}}{\text{distance of sun}}$$

$\therefore$ distance of sun $= \dfrac{4000 \text{ miles}}{\sin BAD}$

BAD is half the measured angle

$$\tfrac{1}{2} \times 17\cdot62'' = 8\cdot81''$$

and $\sin 8\cdot81'' = \cdot0000427$.

Hence distance of sun

$$= \frac{4000 \text{ miles}}{\cdot0000427}$$

$$= \text{about 93 million miles.}$$

7-2

Half the measured angle is called the parallax of the sun. We say the daily solar parallax at the equator is 8·81″. It is this angle which is given in reference books, because it is the angle used in calculations. The measurements need not be made at the equator; but wherever they are made we have to be careful to get the correct base line. We have to be careful also to eliminate the effect of the earth's revolution round the sun and the sun's rotation.

The problem of finding the distance of a star is essentially the same; but it is a far more difficult problem. Attempts were made to find star distances using the diameter of the earth as base. Such attempts were complete failures so far as finding the distances was concerned; because the base was far too small. The biggest base available was the width of the earth's orbit. Instead of waiting 12 hours for a half-turn of the earth on its axis the observers waited six months for a half-revolution of the earth about the sun. Even then the largest star parallax measured is only ¾ of a second.

Without going through the argument again—it is the same as for finding the sun's distance—the distance of the nearest star (α Centauri) is:

$$\frac{93{,}000{,}000}{\sin \cdot 75''} \text{ miles}$$

$$= \frac{93{,}000{,}000}{\cdot 0000036} \text{ miles}$$

$$= \text{about } 26 \times 10^{12} \text{ miles}$$

$$= \text{about } 26 \text{ billion miles.}$$

A unit called a parsec is sometimes used in expressing the

distances of stars. A parsec is the distance of a star whose annual parallax is 1 second. That is:

$$\frac{93,000,000}{\sin 1''} \text{ miles}$$

$$=\frac{93,000,000}{\cdot 0000048} \text{ miles}$$

$$=\text{about 19 billion miles.}$$

It will be seen that the arithmetic of these calculations is easy enough. It is the observations that are difficult. The small star movement that has to be measured is a complicated thing; there is a forward movement of the star as well as the to and fro movement, and the two have to be disentangled before the parallax can be used. We are able to disentangle the two because the forward movement is in a straight line, whereas the parallactic movement is elliptical.

To return to the trigonometrical ratios, there are several interesting points about them.

Tan $Q=\frac{a}{b}$. As the angle gets smaller a gets shorter and shorter and so tan Q gets less and less. Tan $1°=\cdot 01746$; tan $1'=\cdot 00029$; tan $0°=0$ (that is, as the angle gets smaller and smaller its tangent gets nearer and nearer to 0). As the angle Q increases towards $90°$ a gets longer and b gets shorter, so that tan Q rapidly increases. Tan $89°$ is $57\cdot 2900$, and tan $89° 59'$ is $3437\cdot 75$. We say that tan $90°$ is infinite.

Sin $Q=\frac{a}{h}$. At $0°$ a approaches 0, so we say sin $0°=0$. At $90°$ a approaches h in length, so we say sin $90°=1$. Cos $0°=1$; cos $90°=0$.

$\mathrm{Sin}\ Q = \dfrac{a}{h}$. But $\dfrac{a}{h}$ also equals $\cos P$ (a is the base for angle P). Hence:

$$\sin Q = \cos\,(90^\circ - Q),$$

and equally

$$\cos Q = \sin\,(90^\circ - Q).$$

For this reason sines and cosines are often given in the same table. $\mathrm{Sin}\ 1^\circ = \cos 89^\circ$, $\sin 2^\circ = \cos 88^\circ$, $\sin 3^\circ = \cos 87^\circ$, and so on.

Sine, cosine and tangent are not independent of each other:

i.
$$\frac{\sin Q}{\cos Q} = \frac{\dfrac{a}{h}}{\dfrac{b}{h}} = \frac{a}{b} = \tan Q$$

ii.
$$h^2 = a^2 + b^2$$

$$\therefore\quad \frac{h^2}{h^2} = \frac{a^2}{h^2} + \frac{b^2}{h^2}$$

$$\therefore\quad 1 = \sin^2 Q + \cos^2 Q.$$

($\mathrm{Sin}^2\ Q$ stands for the square of $\sin Q$, to distinguish it from $\sin Q^2$, which is the sine of Q^2.)

We can write this relation:

$$\sin^2 A = 1 - \cos^2 A,$$

or

$$\sin A = \sqrt{1 - \cos^2 A}.$$

We can also write it:

$$\cos^2 A = 1 - \sin^2 A,$$

or

$$\cos A = \sqrt{1 - \sin^2 A}.$$

VII

Weights and Measures

ON March 17, 1791 a report—perhaps the most unscientific scientific report ever published—was presented to the French National Assembly. This report proposed the adoption of a decimal system of weights and measures based on the measurement of a quadrant of the earth's circumference along the meridian of Paris. The meridian was measured, and the metric system based on it was finally enforced in France in 1840. Since then the metric system has been adopted by numerous countries for commercial purposes, and almost universally for scientific purposes. The metric system is permissive in Great Britain; but in spite of the earnest efforts of many reformers it has not been enforced to the exclusion of other measures.

One might have thought that a scientific commission would begin, logically, by considering what a unit is. They would have seen that it is a convenient amount of the quantity to be measured, and they would have gone on to find what is the most convenient amount. They might also have considered that the units they devised would be used for two very different purposes—scientific and commercial— and that the needs of scientists are not necessarily the same as those of merchants.

But the commission seems to have been obsessed with the idea of relating the unit of length to some natural standard, and they chose for the unit of length the ten-millionth part of the quadricircumference of the earth measured along the meridian of Paris. The measurement was a matter of great difficulty; it could only be approximate; and there is no certainty that the distance measured is constant. The metre is just as much an arbitrary length as the yard is; its connexion with the quadricircumference of the earth has little scientific value—it enables us to give one measurement in round numbers, and that is all. Indeed a more recent measurement gives the length of the meridian as 10,002,100 metres—so that even the original pseudo-scientific basis of the metre has disappeared.

The other measures were linked up with the metre. For measurements of volume and capacity the litre was a cubic decimetre. The gram was the weight of a cubic centimetre of distilled water at 4° centigrade. The franc was to be 5 grams of gold. All of which had just the right touch of prettiness to appeal to the intelligentsia of all times and all nations; and all rather absurd. The attempt to link the value of the franc to the comparatively stable circumference of the earth sounds, in these fluctuating days, tragically comic.

It is doubtful whether the metric system would have survived but for several pieces of quite unmerited luck. The ten-millionth part of the quadricircumference of the earth turned out to be not greatly different from our convenient yard; if it had been a little longer half the shop-girls in France would have got neuritis in their arms, and their young men would have risen in rebellion. The kilogram, the commercial unit of weight, turned out to be just about twice as big as was needed; and it was instantly halved, so that people could still buy and sell in pounds or pfunds or livres.

The yard is a convenient stretch of the arms. That was how it became a unit of length, and why it remains a unit of length. The metre is a little too long, but the difference is not so great as to make it impracticable as a commercial unit. For some purposes the inch is a better unit than the centimetre. The inch is readily divisible into tenths, and we can estimate hundredths by eye; that is the closest approximation it is possible to get by eye. The metric system gives us by eye estimation nothing closer than the millimetre. We have to remember—what is very often forgotten—that scientific measurements are made in one unit only. The chief advantage claimed for the metric system, that it is completely decimal, is quite illusory. Tenths and hundredths of an inch are just as much decimal as tenths and hundredths of a centimetre. For scientific purposes the comparison is not between a decimal system and a non-decimal system, but between one unit and another unit. English scientists have accepted the metric system for the sake of uniformity, but it has no special advantages over a decimal system based on the inch or yard, and the pound weight.

When we see our measures of length set out in school arithmetics as leagues, miles, furlongs, poles, yards, feet and inches, they do seem rather formidable. But outside school no one ever uses them like that. We do not, for example, measure in yards, feet and inches; we measure in feet and inches when that sort of measurement is indicated; or we measure in yards and half and quarter yards when we are selling materials. The shop-girl stretches a length of material across the convenient measured yard, without being worried by any textbook considerations, or by the extra three and a bit inches inflicted on metrical shop-girls under the pretence of uniting scientific and commercial needs.

The furlong is used by farmers who find it a convenient measure. No one else need bother about it at all; and I doubt whether, outside school, it has ever discommoded anyone. The pole also is retained by people who find it a convenient measure; outside school it is only used when it is convenient; anyone who finds it inconvenient can avoid it with the utmost ease.

For measuring cloth materials we need the single unit of a yard. Even in school arithmetic we do not find anything so absurd as measurements of cloth in miles, furlongs, poles; though we do find measurements given in yards, feet and inches. When we consider that only a single unit is used, all the arguments against the yard, based on its connexion with other units, fall to the ground. The argument for the yard remains. It is a convenient length, based on human necessities. Whereas no one has yet explained what connexion there is between the ten-millionth part (approximately) of a quadrant of the earth's circumference (measured along the meridian of Paris) and the amount of material needed for a dress.

It is really extraordinary how the logical English method of choosing a convenient length as a unit works out in practice. We measure the heights of people in feet and inches to the nearest quarter of an inch. A quarter of an inch is the nearest we can get to the rather variable height, and it is a better approximation than the nearest centimetre. The number of feet in the height gives a first rough approximation to the height. "Four feet", "five feet", "six feet", each says a definite thing which is readily understandable. The idea of a six foot man is a very clear conception to most people. "Five foot eleven" is a little under, "six foot one" is a little over. The corresponding idea of a two metre man is a poor thing by comparison—people of that height are very exceptional, so exceptional that the measure is practically useless. This is not accidental, and it should not be regarded as surprising. The yard is related to the proportioned human body, and it is only to be expected that it should be a suitable unit for body measurements. Whereas no one should expect the ten-millionth part (approximate) of the quadricircumference of the earth (measured along the meridian of Paris) to have any such relation.

The foot-rule is a very convenient length; well, of course it is—that is why it came into use. The best the metric system can do is to give us the two-decimetre rule.

I am not at all impressed by arguments based on the difficulty of multiplying and dividing lengths expressed in miles, furlongs, poles, and so on. Outside school arithmetic books these things do not happen. I defy anyone to attach a probable physical meaning to such an exercise as: 3 miles 5 furlongs 19 poles 1 yard 1 foot 7 inches × 37. If we were to find Chinese children working such exercises we should think "Another Chinese eccentricity!" But alas, familiarity has bred, not contempt, but toleration. Decimalists would have us abolish the convenience because of the absurdity; it seems more sensible to abolish the absurdity and keep the convenience.

The pound is a convenient weight; again that is why we have it as a unit. It is the amount of tea or sugar or rice we want to buy. I have already pointed out that in metric countries the kilogram is usually halved in order to get a more convenient unit. Below the pound we use half and quarter pounds, and so on. The continued halving is exactly what is needed in weighing. In the metric system we get over the difficulty introduced by the ten, by using two weights for the 2-unit—10, 5, 2, 2, 1, ·5, ·2, ·2, ·1, and so on.

Stones, half-stones and quarter-stones introduce no special difficulties. There is no doubt that a stone is a convenient amount for certain commodities; indeed I have seen in Denmark potatoes offered at so much per 3½ kilos which looked like an imitation of the half-stone. For the weights of people stones and pounds are probably as convenient as any weights could be. Eight stones, nine stones, ten stones, etc. have the same sort of relation to weights that feet have to heights; each is a definite standard weight. A pound is as close as one can get to the variable weight of the human body.

A hundredweight is about as heavy a weight as a man should be asked to carry, and a ton is a cart load. A thousand kilogrammes happens to be about 2205 pounds, so that it is very little different from the ton—another piece of luck for the metric system.

The various arguments that are advanced in favour of a decimal coinage are based largely on the idea of paper calculations. The other aspect of the matter is usually left out of account. The superiority of the shilling over corresponding decimal coins lies in its divisibility. 10 is divisible by 5 and 2; 12 is divisible by 6, 4, 3 and 2. We have to keep in mind the tendency of most people to calculate in twos. We have 10, 5 and then 2½ which is not in consonance with the decimal system. Or we have 12, 6, 3 and then 1½ which is in consonance with our system; and we can also have 8, 4, 2, 1, ½, ¼.

Below the penny we have division into halves and quarters, which is all that is required at this stage. Above the shilling we introduce the factor 5 so as to give as great a range of divisibility as possible. Indeed if we compare the complete range of factors of 960 (the number of farthings in a pound), and 1000 (the nearest decimal number) we have the following ranges of factors:

960: 480, 320, 240, 192, 160, 120, 96, 80, 64, 60, 48, 40, 32, 30, 24, 20, 16, 15, 12, 10, 8, 6, 5, 4, 3, 2, 1.

1000: 500, 250, 200, 125, 100, 50, 40, 25, 20, 10, 8, 5, 4, 2, 1.

Thus 960 is exactly divisible in 27 different ways, and 1000 in 15 different ways.

In practice the so-called decimal systems of money are actually centesimal systems. The tenth-unit may exist as a coin, but prices are given in the larger unit and hundredths of this unit. Prices in dollars for instance are given as $3·65, $19·70, etc. We have to make up to a hundred and not to ten. In the English coinage the shilling is a definite landmark. We have to make up to twelve pence only, not to a hundred. The advantage of this may be seen in the ease and certainty with which an English railway clerk, for example, manipulates change.

Many countries have a coin approaching the shilling in value, and this is commonly regarded as a desirable monetary unit. One of the decimal coinages proposed for this country is:

$$1000 \text{ mils} = 100 \text{ cents} = 10 \text{ florins} = £1,$$

but this would entail the abandonment of the shilling as a unit. This would be so stupid a thing to do that it hardly seems possible.

The French franc was originally a coin of the useful shilling class. Its present position should be a warning that decimalization is not a panacea. It is as though we we to calculate all our prices in units of three-halfpence. A wage of £3. 10s. would be inconveniently expressed as 560 of hese units; the cost of a house at £950 would be 152,000.

Oddly enough the florin, the one decimal part of our coinage, and the coin which it is proposed to retain in a completely decimal system, has never been popular. Most people would be glad to see it abolished; two separate shillings would do equally well, and the confusion with the half-crown would end. It is a sign of its unpopularity that it has never had a slang name; it might be worth consideration whether any coins that have not been honoured with slang names should be abolished.

It is odd too that quotations of centesimal currencies end by reverting to the instinctive $\frac{1}{2}$, $\frac{1}{4}$, $\frac{1}{8}$, $\frac{1}{16}$, $\frac{1}{32}$. I see the franc quoted to-day, for example, at $178\frac{13}{16}$. What has become of the centesimal fragments?

The sovereign, or one pound note, stands out as the aristocrat of monetary units. It provides an easy standard of values for such things as wages, the prices of household furniture, and so on. A hundred pounds, a thousand pounds, a million pounds, and a hundred millions, are all admirable units for different purposes. The prices of small houses we reckon in hundreds of pounds, of larger houses in thousands, the national income in hundreds of millions, and so on through a host of everyday affairs.

The argument is sometimes put forward that if we had a decimal coinage prices could be fixed more closely. For example—this is one of the stock examples—tobacco manufacturers might sell us ten cigarettes for 6·9 pence, where they could not reduce the price to $6\frac{1}{2}d$. Let us see where that argument takes us.

In what system of coinage would such a price be possible? Certainly not with the mil, which is about a quarter of a penny and not a tenth. No, the monetary unit would have to be a penny or tenpence; we should have to invent a new coinage to make the reduction possible. And does anyone really want a coin that is worth about half a farthing? In many parts of the country the farthing has almost completely dropped out of use, so there can be little general desire for a still smaller coin. Besides, if such a coin were wanted we could add a half-farthing or a third-farthing to our present coinage; indeed third-farthings are coined.

For what it is worth the argument is an argument in favour of a smaller coin than the farthing, which no one in this country really wants. Its purchasing power would be negligible. We can do all that we need in the direction of derisory reduction of prices with our "one and eleven three".

The guinea has been ridiculed by monetary reformers, but I for one would be sorry to see it go. I have always regarded it as a very handsome gesture on the part of commerce toward the arts and the professions. I should be sorry to have my payments reduced from guineas to pounds, even though every guinea I receive must be a stab through the heart of a pedant.

One would have thought—anyone might be excused for thinking—that decimalists would have wanted to reform the reckoning of time. Surely it should be:

10 seconds = 1 minute;   10 minutes = 1 hour;
10 hours   = 1 day;   10 days   = 1 week;
                  10 weeks = 1 year.

Alas, time defies the pedants. The year is so fantastic a thing —365·2422 mean solar days—that we can only approximate to it with a careful arrangement of leap years. Hours, minutes and seconds are so firmly established as convenient units that there is no doing anything about them. The most that pedantry can do is the twenty-four hour clock, and that we have so far resisted.

Even the reform of the calendar—the touchstone of the pedant—makes little headway. We still have months of different lengths, we still have a variable Easter, and Christmas progressing through the days of the week, we still call the twelfth month the tenth month. The calendar has all its old desperate character—so that our festivals fall oddly and no two years are alike. Which, as every reformer must agree, is a very shocking state of affairs.

The Poet Laureate recently stated that Britons are slaves to their coinage and to their weights and measures. I cannot help thinking that if he will consider the broad humanity of our units, and their relations to human requirements, he will come to see that it is the decimalists who are slaves to a single unresilient number, and that it is we who are logical and free. If his statement is right why does he not—why do not all decimalists—advocate the ten day week that was tried in Russia? If we are not to relate our measures to human requirements why relate the measure of the days to them?

VIII

The Delusive Average

AVERAGING is perhaps the best known of all the secondary arithmetical processes. The method is so simple: you merely add a few quantities and divide the sum by the number of quantities. Even textbook writers do not quite succeed in making the process difficult. They do their worst, but the method remains obstinately simple.

Perhaps because of its simplicity, and the nice air of mathematical certitude about the result—an appearance of exactness obtained by the mere process of dividing—averages have an almost universal appeal. Averaging is the most popular of all arithmetical processes. And the most delusive.

In the physical sciences an average is the mean of several nearly equal results. Suppose three measurements have been made of the same quantity: 13·670, 13·668, 13·665 (the particular unit is immaterial). The average is:

$$\frac{13·670 + 13·668 + 13·665}{3}$$
$$= 13·667\tfrac{2}{3}.$$

This average should be stated as 13·668 and not as 13·667666.... Anything beyond the third decimal place would be the merest pretence, because none of the measurements have been made closer than that.

The advantage of such an average is that we know it to be nearer the truth than the most divergent of the actual measurements. Now we have no means of knowing which of the measurements is farthest from the truth—each is an attempt at the greatest possible accuracy. All three measurements, for all we know, may be too small; if that is the case then 13·665 is the most divergent result and 13·668 (which also is less than the truth) is nearer the truth than this. Or all three measurements may be too great, in which case 13·670 is the most divergent measurement; and 13·668 is nearer the truth than this. And so on with other possibilities.

In finding an average of this kind it would be absurd to take the mean of such results as:

$$15·762, \quad 14·770, \quad 18·768.$$

The average would be meaningless because the second figure is in doubt.

Sometimes one sees examples like:

$$·000017651, \quad ·000017543, \quad ·000016875,$$

where the average is given as ·000017356. This average is almost as absurd as the previous one. The last two figures in the average are meaningless because the figure before them is in doubt. Cutting out pretence the average is ·000017.

Averages are obtained by division and one always has to be extremely careful about results obtained by division. It is so easy to go on dividing and adding figure after figure far beyond the point where meaning ceases. When I see in a reference book that the semi-major axis of the orbit of Mercury is ·3870986 times that of the earth I cannot help wondering where the number came from and how many of the figures are reliable. And when I see in a newspaper that Captain Eyston raced across a measured mile in 10·04

seconds, which is an average speed of 358·57 miles per hour,
I know very well that someone performed the calculation:

$$
\begin{array}{r}
358\cdot56 \\
\overline{1004)\;360000} \\
3012 \\
\hline
5880 \\
5020 \\
\hline
8600 \\
8032 \\
\hline
5680 \\
5020 \\
\hline
6600 \\
6024 \\
\hline
576
\end{array}
$$

358·57 is a slightly closer approximation than 358·56. An
error of 1 in the last figure of 10·04 would make a difference
of about 1 part in 1000. 358·57 ÷ 1000 = ·36, so that the
possible error affects the first decimal place. A correct
statement of the speed would be 358 m.p.h., or 358·6 m.p.h.,
with the last figure doubtful. I have seen recently an account
of a method of timing aeroplanes which are moving at a
speed of 400 miles per hour. The timing can be done to the
twentieth of a second.

$$400 \text{ m.p.h.} = 2 \text{ miles in } \tfrac{3600}{200} \text{ sec.}$$
$$= 2 \text{ miles in } 18 \text{ sec.}$$

$\frac{1}{20}$ second is $\frac{1}{360}$ of 18 seconds; hence the range of error over
a measured length of 2 miles is 1 part in 360, or about 1 m.p.h.
in 400 m.p.h. The account stated correctly that the error was

8-2

not more than ½ m.p.h. Nevertheless the record speed for 3 kilometres (=about 1⅞ ml.) was given as 379·644 m.p.h. It is plain that the last two figures at least are pretence. A better statement would be 379½ m.p.h., and even then the ½ is dubious.

It may have been noticed that in the examples given above the quantities averaged have the same degree of stated accuracy. Children are sometimes asked to find the average of such quantities as:

$$33·05 \text{ cm.,} \quad ·675 \text{ cm.,} \quad 367 \text{ cm.,} \quad 0·00008 \text{ cm.}$$

One could add the four lengths and divide by 4. But what precise meaning can be attached to such an average eludes me; one of the four lengths is less than the probable error in each of the other quantities.

But look again at the first example:

$$13·670, \quad 13·668, \quad 13·665.$$

What precisely is it that we are averaging? There can be no reasonable doubt about the first four figures; they are 13·67. When we try to add a fifth figure to 13·66 we are actually averaging 10, 8 and 5 in order to find this figure.

$$\tfrac{1}{3}(10+8+5) = 7\tfrac{2}{3},$$

and we get 13·667⅔ or 13·668. Now these three numbers are far from being nearly equal. Their divergence from each other shows that measurement to the fifth figure has not been achieved; and therefore the fifth figure should not be added. We can be reasonably sure only as far as 13·67.

And so we come to the rather startling conclusion that the mathematical average does no more than add a bogus final figure to an otherwise accurate result. It can do much worse things than that; that appears to be the best that it can do.

A rather absurd example of this kind of bogus final figure occurred in the stated height of Mount Everest. For years this was given in reference books as 29,002 feet; this (although the fact was not usually stated) was the average of a number of inaccurate measurements. The height is now given as 29,141 feet. Alas poor Two! Poor Two, with its nice air of mathematical precision that we had all so fondly doted on. What a fate—to be swallowed up by 139 other dubious feet! 29,002 was almost as well known as 1066: but who is going to remember 29,141?

There is another kind of mathematical average which is not open to this objection. We talk about the average density of the earth, and this meaning of average is different from the average of several measurements. It may of course be found as the average of several measurements; but the result of one good measurement would still be called the average density of the earth. It is the density of the earth taken as a whole—as if it were of the same density throughout. The averaging comes in in spreading the whole measured density throughout the whole volume. Such an average applies to the whole mass only; there is no pretence that it indicates the density of any particular part. But if we find that the surface density is lower than the average we can say that the interior density is above the average. With proper safeguards averages of the kind we have been considering can be used with confidence. We can use the average density in finding the mass of the whole earth, or in comparing one planet with another.

The average speed of a train on a journey is another example of an equalizing average. The actual speed may be varying almost continuously; at the start it works up from zero to a high speed, and at the end it slows down again to zero. The average speed is a summary of the whole journey; it does not give the speed at any particular part. Such averages are useful as indicators for other similar journeys.

It is an interesting point that the average of two speeds is meaningless, except in the case where the speeds are maintained for equal times. Captain Eyston's average speed for the double run (northward and southward) was not the mean of the two speeds but the total distance (2 miles) divided by the total time, or, what amounts to the same thing, 1 mile divided by the average time. The difference is small, because the times were nearly equal. But it can readily be shown that to travel equal distances at faster and slower speeds always takes longer than to travel the whole distance at the mean of the two speeds. A famous example is: a boat rowed a mile upstream and a mile downstream with equal energy. The total time is greater than to row two miles in still water with the same energy. Thus, with a speed of 3 m.p.h. in still water and a current of 2 m.p.h. the speed upstream is $3 - 2 = 1$ m.p.h. and downstream $3 + 2 = 5$ m.p.h. Thus the total time is: 60 min. + 12 min. = 72 min. The time to do two miles in still water is: 40 min. The reason for the difference should be apparent: in the up-and-down movements the boat is rowed for a longer time at the slower speed.

In the Morley-Michelson experiment the speed of light was timed up and down a supposed stream of ether, and across the stream. Careful experiment showed that there is no difference at all. This was one of the bases from which the Einstein theory was reasoned out. It seemed decisive against a material stream of ether.

Greenwich mean time is another equalizing average. The solar day varies in length throughout the year because of the varying speed of the earth in its orbit; it is sometimes longer than 24 hours and sometimes shorter. The 24 hour day is the average length of the day throughout the year. The average day is a most valuable average. Without it we should have days of differing length, clocks would have to be adjusted even though they worked with extreme accuracy, and odd minutes would accumulate that were not reckoned in the 24 hours. A properly adjusted sundial shows solar time. Ignoring the changed hour of "summer time", sundial noon varies from the noon of Greenwich mean time by amounts varying from zero up to about twenty minutes.

The mathematical idea of averages has been extended in many directions, often without the safeguards which usually limit its powers of pretence in the physical sciences. Things are averaged in the same sort of way as in physics and chemistry, though they may be far from being equal. We find average ages, average incomes, average heights, average cricket scores, and so on through a host of things measurable or not measurable. The economist produces the average economic man, and the psychologist averages human faculties.

The "ordinary or average man" is a real average. We imagine the peculiarities and abilities of many people spread out evenly, and we thus produce the average man—of medium height and weight, and without any striking qualities. But the average Englishman—or Frenchman or Dutchman—is produced in a most peculiar way. Suppose we want to produce the average Englishman—we take all the qualities that are exaggerated in certain Englishmen, we add them all together, and that is your average Englishman. We have forgotten—what every schoolboy knows—that you cannot have an average without dividing. Thus Mr Shaw: "the hysterical, nonsense-crammed, fact-proof, truth-terrified, unballasted sport of all the bogey panics and all the silly enthusiasms that now calls itself 'God's English man'." Surely a little exercise in division was called for, even though the result might have been less rhapsodical.

Some equalizing averages have a limited use, but they should all be regarded with the gravest suspicion, lest more be read into them than is warranted by the facts. The average of £50 a year and £50,000 a year is:

$$\pounds\,\frac{50 + 50,000}{2} = \pounds 25,025 \text{ a year.}$$

But that is a mathematical process applied quite wrongly, and the result is an absurdity. One man with £50 a year and another with £50,000 a year are not at all the same thing as two men with £25,025 a year each.

I have purposely chosen an extreme case where the crudity and absurdity are obvious. But here is a case, just as extreme, where nature seems to have done the averaging. *The Phenological Report* 1933 ends with the following: "The isophanes are therefore pushed south in the north whilst those in the south are further north than usual, thus curiously making for the British Isles a normal year!" That is the "average pretence" in a sentence. Since the north is abnormal in one way, and the south in the opposite way, therefore the country as a whole is normal—which it isn't!

When there are only two or three things being averaged the crudity and falsity of the result are often obvious, especially when they are wide apart in value. But it is usual to include large numbers of quantities in such averages; the crudities are thus disguised—but they remain. It is the disguise that constitutes the danger. A statement of average income makes no difference between a country where much wealth is accumulated in a few hands whilst the bulk of the people are in poverty, and a country where there is a general fairly high level of wealth. We can get a better result by averaging at different levels—the average of high incomes, the average of medium incomes, and the average of low incomes. Indeed it should be clear that the further we get away from the general average the nearer we get to the truth. It is at least doubtful whether an average ever adds anything to the truth, though it may make certain aspects of it more readily comprehensible.

In the case of incomes we can get a more satisfactory result by departing from the average and stating the number of incomes within certain ranges: below the poverty line, comfortable incomes, large incomes, and very large incomes.

We usually have little to guide us, beyond observation and common sense, as to whether averages are fair averages or not. The average yearly consumption of tea per head in the British Isles is about 10 pounds. That is a not unfair average because we know that almost everyone in this country drinks tea, and that even excessive tea drinkers do not drink so very much more than moderate tea drinkers. (It is not as if some consumed 50,000 pounds of tea per annum, whilst others consumed only a few ounces.) The average consumption of bread and flour also are not unfair averages. But the average consumption of caviar in England per head would hardly have much meaning or value.

We are given from time to time the average rainfall for each month and for the whole year—the various results are added up and nicely divided out to two places of decimals. If we go on averaging for enough years we shall reach a point when a year of the heaviest possible rainfall can have no effect on the average. If our descendants have rain on every day of the year they will be able to comfort themselves with the fact that the average rainfall (to two places of decimals) cannot be affected. If the average were taken across a long period of rather damp years it would be a good many years before a period of rather dry years would have any very appreciable effect on the average.

We cannot always be sure where the meaning in an average ends and where nonsense begins. The average yearly rainfall throughout England has some sort of significance, because the rainfall does not vary greatly from place to place. But the average rainfall for the whole of Australia approaches sheer nonsense. We have to examine every average and consider it on its merits; and suspicion is a virtue.

There is one set of averages that should be scrutinized with extreme care. Psychologists have a bad habit of expressing their results in mathematical form, and so giving them a quite spurious appearance of exactness. And when they average their "mental quotients", applying a dubious arithmetical process to a mass of dubious results, one does not merely suspect the result.

Psychologists profess to measure general intelligence (having previously disclaimed any knowledge of what general intelligence is) by means of the answers to a number of rather absurd, and sometimes rather amusing, questions called "intelligence tests". A not unfair example of such questions is "If a white cow gives white milk what is the colour of the milk given by a black cow?" (*Master*: "Why on earth did you say 'green'?" *The Boy* (a farmer's son): "Well, I thought you wanted something different; and it does look pretty green sometimes before it's siled." Intelligence—nil.) The arithmetic comes in in assessing the number of marks gained by the child and in dividing it by the possible number of marks—to as many figures as the conscience of the psychologist permits. When the number of marks has been manipulated into an intelligence quotient it acquires an almost mystic significance. We can proceed, with sufficient expenditure of time and energy, to find the average mental quotient for a whole school, or a whole city, or even for a whole country. Arithmetic, what crimes are committed in thy name! On the whole it would be well to keep firmly in one's mind the exact origin of the psychologist's numbers and formulae, and to regard his averages with even more suspicion than usual.

I once propounded the conundrum "What is the average of Shakespeare and Lenin and Einstein?" My son replied instantly "H. G. Wells". But I have no doubt a psychologist would have given an intelligence test to all three, labelled each with a number, and found the average to two places of decimals.

IX

Approximations

To hear some people talk one might imagine that a mathematical approximation is a sloppy sort of thing. So far from that being true a good approximation is an attempt to state the truth as exactly as we can.

In arithmetic we are dealing with three kinds of quantities. In the first place we have quantities that can be counted. We can count the number of chairs in a room, the number of bricks used in building a house, the number of books in a house, the population of a city, or the revenue of a country. Such things must have two qualities to make them countable: they must be separate units, and they must be alike. The likeness need not extend very far; the books may vary greatly one from another, but they must be sufficiently alike for each to be called a book. Quantities that can be counted are sometimes called discrete quantities. It is not to the point that they may be difficult to count. It would be tedious to count the number of grains in a stone of rice; nevertheless they could be counted.

The second kind of quantity is continuous and not discrete. Length is a good example of a continuous quantity. When we are measuring a length we are doing something very different from counting a sum of money or a number of books. We are so familiar with the use of rulers that we are apt to forget what we are actually doing in measuring a line. We are not counting one, two, three, and so on, but comparing the length of a line with some standard length—an inch, or a yard, or a metre. It is only when there is no ruler handy that we feel the need of a standard length. We may find a length by pacing—we are comparing the length with the length of a pace. We may measure a room roughly with a walking stick—we are comparing its length and width with the length of the stick.

Other continuous quantities are areas, volumes, masses, weights and time.

The third kind of quantity is abstract numbers. We usually learn about numbers by reference to the counting of some kind of object. Afterwards we let the idea of the objects fade away and retain the numbers as abstract ideas. In arithmetic we often deal with numbers merely as numbers and without reference to any physical reality.

Each of these three kinds of quantities is measured in its own way. Discrete quantities are countable. The huge national revenue can be counted to a single penny. Transactions at the Bankers' Clearing House, amounting to perhaps forty thousand million pounds in a year, can equally be accounted for to a penny. Census returns are made to a single unit. And so on. There are, however, many discrete quantities whose number can only be estimated. Atoms and molecules obviously come into this class because of their diminutive size and vast numbers. Census returns are only exact in so far as the forms are filled in correctly, and even then the exactness has disappeared long before it is possible to announce them.

The number of people admitted, say, to Kew Gardens can be found from the number recorded at the turnstiles. But the number of people in an unenclosed crowd can only be estimated. The traditional Eastern potentate might enclose them and insist on an exact count, but the only estimate possible to western Europeans is a comparison with the counted crowds at (say) football matches.

We can sometimes reduce the labour of estimating large numbers of small objects. Suppose again—we have already supposed this—that we want to find the number of grains in a stone of rice. We might go to the excessive trouble of counting all the grains in a stone of rice; but we can reduce the trouble considerably. We might carefully weigh a quarter of an ounce of rice and count the number of grains in this quarter of an ounce. We then multiply the counted number of grains by $4 \times 16 \times 14 = 896$, and so find the number of grains to within as close an approximation as we could possibly want. The actual counting is reduced to about one nine-hundredth of the total number.

In estimating the number of stars that can be seen through a telescope we can use a similar method. We mark off a definite portion of the sky and carefully count the number of stars in this part. We multiply the number counted by the number of times the measured part is contained in the sky. We should have to make allowance of course for parts of the sky where there were fewer stars. But this method would give a rough approximation to the total number of visible stars.

In dealing with continuous quantities we have a different problem. We are not counting—one, two, three, and so on— but placing the quantity to be measured side by side with a standard quantity. Our rulers for measuring lengths are marked off in inches, and we do count the inches—one, two, three, and so on; this rather disguises what we are actually doing. One inch, two inches, three inches, and so on, are all repetitions of our standard length with which we are comparing the length to be measured. We are not counting

the number of inches in the same sense that we are counting
the number of chairs in a room. There may be six chairs;
it would be the merest quibble to call a broken chair a
fraction of a chair; either it is a chair in the count, or it isn't;
it is certainly not a fraction of a chair in the sense that we
have a fraction of an inch. When we say that a line is
3½ inches long we do not mean that it is exactly 3½ inches
long, without the possibility of the most infinitesimal error.
In the finest of rulers the dividing marks have an appreciable
width; under a microscope a line, however carefully drawn,
has rough ends, so that one hardly knows exactly where to
begin measuring; and even if the beginning of the line is
exactly right we can never be quite sure that the length we
read for the end is not slightly out. With extreme care we
might measure a line to the nearest ten-thousandth of an
inch; but even if we measured more closely, say to the
millionth of an inch, we could not possibly be sure that there
was not a still smaller error. Sooner or later we reach a limit
to the possibility of exact measurement; and beyond that
there is no certainty. That is why all measurements of
continuous quantities are in their very nature approximate.
Even if by chance we did happen to get an exact measurement we should have no means of knowing that it was exact;
there might be an error beyond our range of measurement.

A small child can with difficulty measure short lengths in
inches to the nearest tenth of an inch; a boy of 15, provided
with a good ruler and estimating the hundredths by eye, can
measure to the hundredth of an inch; with a vernier he can
measure to the thousandth of an inch; and there are gauges
that enable one to make even finer measurements. But
sooner or later there is an end to what is possible.

When we give the length of a line we should always state

exactly what we mean; that is the true meaning of approximation. 6 inches long is a vague measurement; it may mean
that we have made a rough estimate of the length—such
an estimate as would be good enough if we were cutting,
say, a length of glass tubing, where a quarter of an inch more
or less did not matter. 6·0 inches is a much more definite
statement. It declares that the length has been measured to
the nearest tenth of an inch; it is not so much as half the
tenth of an inch out. That is to say it is less than 6·05 inches
long and more than 5·95 inches long; these are the extreme
limits of error. If the line were a little more than 6·05 inches
we should give its length as 6·1 inches (true to two figures, or
to the nearest tenth of an inch); if it is less than 5·95 inches
we should give the length as 5·9 inches (again true to two
figures).

6·00 inches means that the line has been measured to the
nearest hundredth of an inch; it is less than 6·005 and greater
than 5·995. 6·000 claims to be a very exact measurement—to
the nearest thousandth of an inch. And so on.

Children are sometimes asked to add up quantities like
this:

$$3·5 \text{ in.} + ·0265 \text{ in.} + 36 \text{ in.} + 14·673 \text{ in.}$$

Presumably they are expected to work the sum like this:

$$
\begin{array}{r}
3·5 \\
·0265 \\
36 \\
\underline{14·673} \\
54·1995 \text{ in.}
\end{array}
$$

Such a sum is ridiculous. In the measurement 3·5 there is a
possible error of ± ·05, so that any figures beyond the first
decimal place are quite unreliable. Even allowing some

degree of accuracy in the measurement 36 in., a correct answer would be 54·2 in.

We sometimes speak of "three figure accuracy", "four figure accuracy", and so on. It is the number of correct figures in a measurement that matters. 3·674 miles is as exact a measurement as 3·674 inches; it is the same measurement on a different scale. The idea of proportion, which runs right through arithmetic, should prepare us for such a result. The difference between the two measurements is, we say, a matter of scale. In the large scale measurement everything is exaggerated: defects in the measuring apparatus are magnified so that possible errors are magnified. When we increase the scale, as a general rule we increase the errors in the same proportion, so that we get the same degree of accuracy.

Many measurements are based on the measurement of angles, which can often be measured with great accuracy.

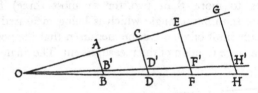

Suppose we measure the angle at O and use this as a basis for finding the lengths AB, CD, EF, GH. A slight error in measuring the angle O would give the length AB' instead of AB, CD' instead of CD, and so on. The triangles OAB, OCD, etc. are all similar triangles, so that corresponding lines are in proportion. The errors BB', DD', FF' and HH' are proportional to the lengths AB, CD, EF and GH. If the error in measuring AB were one-thousandth, so also would be the errors in measuring CD, EF and GH.

Here is another way of looking at the matter. Suppose the length of a line close at hand is being measured, the width of the angle to be measured is comparatively great. As one's distance from the line increases the size of the angle decreases. Suppose we can measure angles with a possible

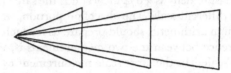

error of 1 minute of arc. In measuring an angle of 10 degrees the error is 1 part in 600; in an angle of 1 degree the error is 1 part in 60, that is ten times as serious as before. So that with increase of distance this element of inaccuracy increases. That is one reason why it is impossible to measure star distances to more than two (or at most three) figure accuracy; the minute angle which is being measured is so close to the limit of what can be measured that the possible error is a large fraction of the measurement. The angle *ACB*

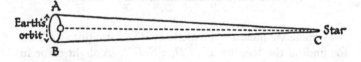

which has to be measured is the same as the angular width of the earth's orbit as seen from the star. For the nearest star the angle is $1\frac{1}{2}$ seconds of arc. To find a comparison for this measurement let us find at what distance a halfpenny (one inch across) must be placed so as to have an angular width of $1\frac{1}{2}$ seconds.

In the diagram $\sin \frac{3}{4}'' = \dfrac{AO}{AC}$, and this is also to equal $\dfrac{\frac{1}{2}\,\text{inch}}{x}$ (x is the distance at which the coin must be placed).

$$x = \frac{\frac{1}{2}\,\text{inch}}{\sin \frac{3}{4}''}$$
$$= \frac{\frac{1}{2}\,\text{inch}}{\cdot0000036}$$
$$= \text{nearly } 140,000 \text{ inches}$$
$$= \text{more than 2 miles.}$$

So that that extraordinary star measurement is equivalent to measuring the width of a halfpenny two miles away.

One other point about the measurement of lines should be noted. We cannot find lengths by measuring angles only. The angles only give us the proportions. When we know the angles of a triangle we know its shape, but not its size. Every measurement of length depends finally upon some directly measured length; and the accuracy of the calculated length can be relied on no further than the direct measurement on which it depends.

Weight is another continuous quantity which can only be measured approximately. For trade purposes we often try to treat weight as if it were discrete. Tea is made up into pound, half-pound, and quarter-pound packets, and these are sold as separate units. So far as the customer and the shopkeeper are concerned the packets of tea are discrete quantities. That is because the approximate weighings have already been done—by the wholesale merchant. His scales give only approximate weighings, and a leaf more or less is neither here nor there. No two packets of tea ever weigh exactly the same—the possibility that they should have exactly the same weight is so remote that it is not worth considering. However small the limits of error we allow there is an infinite number of possible gradations. And if by chance two packets did weigh exactly the same we should have no means of knowing.

Minute discrepancies in weights are of no importance commercially; we mention them to illustrate the approximate nature of all weighing—a matter which is of extreme importance in scientific weighings. Commercial weighings also illustrate another important point. A small inaccuracy which would be tolerated in the weighing of a ton of coal, and which is indeed inseparable from such weighing, would not be tolerated in the weighing of a pound of tea. The weight of a small knob of coal is neither here nor there in the weight of a ton; it is equivalent to a weight 1/2240th as great in a pound.

In scientific work the exact degree of approximation in weighing must be stated, just as it is for measurements of length. 2·0 gm. is true to the nearest tenth of a gram (a rough weighing); 2·00 gm. is true to the nearest hundredth of a gram, and so on.

All the units which depend on length and weight are continuous quantities, and can only be measured approximately. In measuring areas there is often a double approximation, since we measure two lengths and multiply them together. We may find the radius of a circle to be 2·37 in. The area is given by $3\frac{1}{7}r^2 = 3\frac{1}{7} \times 2·37^2$ sq. in.

$$
\begin{array}{r}
2·37 \\
2·37 \\
\hline
4·74 \\
·711 \\
·1659 \\
\hline
5·6169
\end{array}
\qquad
\begin{array}{r}
5·6169 \\
3\frac{1}{7} \\
\hline
16·8507 \\
·8024 \\
\hline
17·6531
\end{array}
$$

It might appear that the area is 17·6531 square inches. But consider the possible errors. 2·37 is true to the nearest hundredth. We know that the true length is less than 2·375 and greater than 2·365:

$$3\tfrac{1}{7} \times 2·375^2 = 17·727678,$$
$$3\tfrac{1}{7} \times 2·365^2 = 17·578707.$$

We know that the correct area lies between these extremes; we cannot be at all sure of anything closer than $17·65 \pm ·07$ (where $\pm ·07$ indicates the possible error). The same kind of precaution is even more necessary in calculating a volume from a measured length.

Density is the amount of material per cubic centimetre or per cubic foot. Thus the density of water is 1 gram per cubic centimetre or 62·3 lb. per cubic foot (for most purposes we say $62\tfrac{1}{4}$ lb. or $62\tfrac{1}{2}$ lb. per cubic foot). Measurements of density therefore are doubly approximate; they depend on two approximations—length and weight.

Time is a continuous quantity which can sometimes be measured with extreme accuracy. We sometimes get an event repeated in the same time over and over again. Thus in timing the swing of a long pendulum we might easily have an error of a tenth of a second. But in timing a thousand swings the error need not be any greater; thus the time of a single swing can be found by division to the ten-thousandth part of a second. Chronometers are made which can be adjusted to give errors considerably less than a second in a day. A second in a day is:

1 part in 24 × 60 × 60,

that is   1 part in 86,400,

so that time measurements can be made very closely.

Measurements of speed and acceleration depend on both time and length measurements, so that the degree of approximation depends on the measurement of length and on the method of timing. Electrical contacts enable the timing to be done very accurately.

The approximations that have been given so far all include a decimal point. In 3·045 we have four figure accuracy; in ·0003045 we equally have four figure accuracy. But what about 3,045,000? Is that measurement true to four figures, or to five figures (the fifth figure just happening to be 0), or is an even closer approximation intended (with two or all three of the end figures happening to be 0)?

There is nothing to tell us. Very often we are left to guess what the answer is. When we say that the distance of the sun is 93,000,000 miles we *know* that only two figure accuracy is intended; if we do not know there is nothing in the number to tell us. There is, however, a way, commonly used by scientists and others, of stating the exact degree of approximation. Numbers are written in such a way that they indicate the approximation. $3,600,000 = 3·6 \times 10^6$. That is, we write the number with its first figure as a units figure, and multiply by the appropriate power of ten (to find the power simply count the number of moves to carry the decimal point from its original position to its place after the units figure). $3·6 \times 10^6$ indicates two figure accuracy. $3·60 \times 10^6$ is the same number measured to three figure accuracy; $3·600 \times 10^6$ indicates four figure accuracy; and so on.

This way of writing numbers is called writing them in standard form. It may be used with any number:

$$67313 = 6·7313 \times 10^4,$$

$$·0027 = 2·7 \times 10^{-3} \quad \text{and so on.}$$

Standard form is useful for writing down cumbrous numbers.

$$1 \text{ light-year} = 5{,}900{,}000{,}000{,}000 \text{ miles}$$
$$= 5 \cdot 9 \times 10^{12} \text{ miles.}$$

The mass of a hydrogen molecule is:

$$\cdot 00000000000000000000000332 \text{ gm.}$$
$$= 3 \cdot 32 \times 10^{-24} \text{ gm.}$$

The power of 10 indicates the scale of the measurement, without our having the trouble of counting the number of noughts.

If we were multiplying or dividing these two numbers it would be a woeful task to set out all the working in full. Standard form enables us to work more expeditiously and certainly:

$$5 \cdot 9 \times 10^{12} \times 3 \cdot 32 \times 10^{-24} = 5 \cdot 9 \times 3 \cdot 32 \times 10^{-12}$$

$$
\begin{array}{r}
3 \cdot 32 \\
5 \cdot 9 \\
\hline
16 \cdot 60 \\
2 \cdot 988 \\
\hline
19 \cdot 588
\end{array}
$$

The product is $1 \cdot 96 \times 10^{-11}$

$$5 \cdot 9 \times 10^{12} \div (3 \cdot 32 \times 10^{-24}) = 5 \cdot 9 \div 3 \cdot 32 \times 10^{36}$$

$$
\begin{array}{r}
1 \cdot 77 \\
332 \overline{)\,590} \\
332 \\
\hline
2580 \\
2324 \\
\hline
2560 \\
2324 \\
\hline
236
\end{array}
$$

The quotient is $1 \cdot 78 \times 10^{36}$

When we come to pure numbers we are not tied down in the same way as with continuous quantities. Approximations in continuous quantities are due to imperfections in the means of measurement. With pure numbers we can go to any degree of approximation we choose, because we depend entirely on calculation and not on measurement. We can say that the circumference of a circle is:

3·14159265358979323846 times the diameter.

But we do not pretend that we could ever arrive at such a result by measurement, or that we could use that number for any circle that we could draw, no matter how fine the instruments. The number arises from a calculation based on the nature of an ideal circle denuded of all the imperfections of materials. A displaced atom might falsify a calculation based on the value of π given above.

The square root of 2 is a definite enough number—it is the number which multiplied by itself gives 2 as a product. It is quite easy to show that it can never be written down exactly as a decimal. Suppose it could be written down, we should have 1·414... say to a million places. The last figure cannot be nought, or else we should have one less place. We multiply this cumbrous number by itself and we get a number running to two million places of decimals; again the last figure is not 0, because it is obtained by squaring a number which is not 0. Thus we get a number which is not 2 exactly, because it has a figure in the two millionth place. Therefore the square root cannot end at the millionth place, or at any other place. We say that $\sqrt{2}$ is incommensurable.

The square root of 2 is 1·41421...

		Difference from 2
$1·4^2$	$= 1·96$	·04
$1·41^2$	$= 1·9881$	·0119
$1·414^2$	$= 1·999396$	·000604
$1·4142^2$	$= 1·99996164$	·00003836

We have the successive approximations to $\sqrt{2}$: 1·4, 1·41, 1·414, 1·4142. The square of each is a closer approximation to 2 than that of the previous number. With time and care we might arrive at a number whose square is one point any number of nines we like. We can get readily enough to within a billionth or a trillionth or a quadrillionth of 2, but we can never get to 2.

The roots of all the numbers are incommensurable with the exception of the few that are perfect squares: 0, 1, 4, 9, 16, 25, 36, 49, 64, 81, 100, and so on. The square numbers get wider and wider apart; the differences are 1, 3, 5, 7, 9, 11, and so on—the sequence of odd numbers. So that the number of perfect squares in a range of say 100 numbers decreases as we advance. There are 11 from 0 to 100; only 4 from 101 to 200 (121, 144, 169, 196), and only 3 from 201 to 300 (225, 256, 289).

e is another incommensurable number; it is the number used in calculating logarithms:

$$e = 1 + \frac{1}{\lfloor 1} + \frac{1}{\lfloor 2} + \frac{1}{\lfloor 3} + \frac{1}{\lfloor 4} + \frac{1}{\lfloor 5} + \text{etc.}$$

The series goes on indefinitely.

$\lfloor 1$ stands for 1, $\lfloor 2$ for 1×2,

$\lfloor 3$ for $1 \times 2 \times 3$, $\lfloor 4$ for $1 \times 2 \times 3 \times 4$,

and so on. We call $\lfloor 4$ "factorial four". Each fraction can be readily obtained from the one before it by dividing by 2 (for $\lfloor 2$), by 3 (for $\lfloor 3$), by 4 (for $\lfloor 4$) and so on. When we reach $\lfloor 10$ each number will be a tenth or less of the previous one, and so we shall soon reach a point where the fractions cease to count. Here is the calculation with the fractions

worked out to ten places. (The first 3 terms are put together):

$$2 \cdot 5$$
$$\cdot 1666666667$$
$$\cdot 0416666667$$
$$\cdot 0083333333$$
$$\cdot 0013888889$$
$$\cdot 0001984127$$
$$\cdot 0000248016$$
$$\cdot 0000027557$$
$$\cdot 0000002756$$
$$\cdot 0000000251$$
$$\cdot 0000000021$$
$$\cdot 0000000002$$
$$\overline{2 \cdot 7182818286}$$

That is the value of e to 9 places of decimals; the last
figure is unreliable. It is the value to a thousandth of a
millionth. We can calculate e to any degree of accuracy we
like. It would be quite easy to add four more figures to each
line and four more lines, and so add four more figures to the
value of e.

π is another incommensurable. Approximate values are
$3\frac{1}{7}$, $3 \cdot 14$, $3 \cdot 1416$ $3 \cdot 14159$ (we use, or should use, whichever
degree of approximation suits our other approximations).
$3\frac{1}{7}$ by the way is a slightly more accurate value than $3 \cdot 14$.
$3\frac{1}{7} = 3 \cdot 14286$, and this differs from $3 \cdot 14159$ by $\cdot 00127$, whereas
$3 \cdot 14$ differs from it by $\cdot 00159$.

There are many series for calculating π. Some of them
converge very slowly—the terms decrease so slowly that you
have to take a great number of them before you get a
reasonably accurate result. One such series is:

$$\tfrac{1}{4}\pi = 1 - \tfrac{1}{3} + \tfrac{1}{5} - \tfrac{1}{7} + \tfrac{1}{9} - \tfrac{1}{11} + \text{etc.}$$

We have to add $1+\frac{1}{5}+\frac{1}{9}+\frac{1}{13}+$ etc., subtract $\frac{1}{3}+\frac{1}{7}+\frac{1}{11}+$ etc., and then multiply by 4. It is an interesting, and exasperating, experience to try to calculate π from that series.

But there are much better series than that for calculating π:

$$\tfrac{1}{4}\pi = \quad \tfrac{1}{2}-\tfrac{1}{3}\left(\tfrac{1}{2}\right)^3+\tfrac{1}{5}\left(\tfrac{1}{2}\right)^5-\tfrac{1}{7}\left(\tfrac{1}{2}\right)^7+\dots$$
$$+\tfrac{1}{3}-\tfrac{1}{3}\left(\tfrac{1}{3}\right)^3+\tfrac{1}{5}\left(\tfrac{1}{3}\right)^5-\tfrac{1}{7}\left(\tfrac{1}{3}\right)^7+\dots.$$

This series is called Euler's series. It converges much more rapidly than the previous one.

Another series of Euler's is:

$$\pi = 2\cdot8\left[1+\tfrac{2}{3}\times\tfrac{2}{100}+\tfrac{2}{3}\times\tfrac{4}{5}\times\left(\tfrac{2}{100}\right)^2+\tfrac{2}{3}\times\tfrac{4}{5}\times\tfrac{6}{7}\left(\tfrac{2}{100}\right)^3+\dots\right]$$
$$+\cdot30336\left[1+\tfrac{2}{3}\times\tfrac{144}{100000}+\tfrac{2}{3}\times\tfrac{4}{5}\times\left(\tfrac{144}{100000}\right)^2+\dots\right].$$

The sines, cosines, and tangents in trigonometrical tables are calculated from series. These series give the angle in radians; that is why this rather extraordinary angle bulks so largely in trigonometry.

Everyone knows that if you step off the radius round the circumference of a circle it goes six times exactly and you get angles of 60°. If instead of stepping off the radius you actually bend it round the curve it will go more than six times—in fact twice $3\frac{1}{7}$ times. With an arc one radius long we get an angle rather less than 60°; this angle is called a radian. The radius goes $3\frac{1}{7}$ times into a half-circle, and so there are $3\frac{1}{7}$ radians in a half-circle or 180° (or more exactly 3·14159 radians = 180°).

$$180° = 3\cdot14159 \text{ radians}$$
$$1° = 3\cdot14159 \div 180 \text{ radians}$$
$$= \cdot017453 \text{ radian}.$$

Hence to change degrees to radians we multiply the number of degrees by ·017453.

There are very elegant and simple series for calculating sines and cosines:

$$\sin \theta = \theta - \frac{\theta^3}{\underline{3}} + \frac{\theta^5}{\underline{5}} - \frac{\theta^7}{\underline{7}} + \text{etc.}$$

(θ is the angle reckoned in radians),

$$\cos \theta = 1 - \frac{\theta^2}{\underline{2}} + \frac{\theta^4}{\underline{4}} - \frac{\theta^6}{\underline{6}} + \text{etc.}$$

Suppose we want to find $\sin 1°$.

$$1° = \cdot 017453 \text{ radian} = \theta,$$

$$\sin 1° = \theta - \frac{\theta^3}{\underline{3}} + \frac{\theta^5}{\underline{5}} - \frac{\theta^7}{\underline{7}} + \ldots$$

$$\theta = \cdot 017453,$$

$$\theta^2 = \cdot 017453^2 = \cdot 0003046,$$

$$\theta^3 = \cdot 017453^3 = \cdot 00000532.$$

If we need further odd powers of θ we multiply $\cdot 00000532$ by $\cdot 0003046$ as often as is necessary. But for $\sin 1°$ we need only:

$$\theta = \cdot 017453; \quad \frac{\theta^3}{\underline{3}} = \cdot 0000009.$$

$$\sin 1° = \cdot 017452.$$

For extreme accuracy we might add $\frac{\theta^5}{\underline{5}}$ and subtract $\frac{\theta^7}{\underline{7}}$, but there are not many purposes for which $\cdot 017452$ would not be more than accurate enough.

When θ is small θ^3 is very small and θ^5 is extremely small. In $\sin 1°$, θ^3 makes a difference of 1 in the sixth decimal place. For angles less than $1°$ the difference would be still more insignificant. Thus we can say that for angles under $1°$,

$$\sin \theta = \theta.$$

When we were calculating the distance of α Centauri we needed sin $\frac{3}{4}''$:

$$1° = \cdot017453 \text{ radian}$$
$$1'' = \cdot017453 \div 3600 \text{ radians}$$
$$\tfrac{3}{4}'' = \tfrac{3}{4} \times \cdot017453 \div 3600 \text{ radians}$$
$$= \cdot00000363 \text{ radian}.$$

Hence   sin $\frac{3}{4}'' = \cdot00000363$.

Now the distance of a star (we have already noticed this)

$$= \frac{\text{radius of earth's orbit}}{\text{sine of angle}}$$

$$= \frac{\text{radius of earth's orbit}}{\text{angle}}$$

That is the distance is inversely proportional to the parallax —half the parallax, double the distance. Arcturus has a parallax of $\cdot024''$ and Capella of $\cdot12''$. The parallax of Arcturus is $\frac{\cdot024}{\cdot12} = \frac{1}{5}$ of that of Capella; that is Arcturus is 5 times as far off as Capella.

The series for sin θ:

$$\sin\theta = \theta - \frac{\theta^3}{\lfloor 3} + \frac{\theta^5}{\lfloor 5} - \frac{\theta^7}{\lfloor 7} + \text{etc.}$$

is very rapidly convergent when θ is small; it converges so rapidly that when the angle is less than $1°$ we need only take the first term. But the series becomes more troublesome as θ becomes larger. When θ equals a radian (that is 1) the series becomes:

$$\sin\theta = 1 - \frac{1}{\lfloor 3} + \frac{1}{\lfloor 5} - \frac{1}{\lfloor 7} + \text{etc.}$$

and beyond a radian the powers of θ get bigger and bigger.

But we need actually go no further than 45° in calculating sines and cosines. We have already noted that:

$$\sin A = \cos (90° - A),$$
and $$\cos A = \sin (90° - A).$$

If we calculate the sines from 0° to 45° these are the same as the cosines from 90° to 45°. And similarly the cosines from 0° to 45° are the same as the sines from 90° to 45°.

In a table of sines and cosines it is usual to give the angles up to 45° on one side, and angles from 90° to 45° on the other side:

	sin	cos	
0°	·0000	1·000	90°
1°	·0175	·9998	89°
2°	·0349	·9994	88°
3°	·0523	·9986	87°
etc.			etc.
to			to
45°	·7071	·7071	45°
	cos	sin	

The angles 1°–45° are read from the top, the sines on the left and cosines on the right. The angles 45°–90° are read from the bottom, the cosines on the left and sines on the right.

X

Multiplication and Division

As a rule division is harder than multiplication. We often resort to artifices in order to avoid awkward divisions.

Suppose we need to find $\frac{1}{\sqrt{2}}$, we could work out $\sqrt{2}$ and then divide 1 by this number. We avoid the difficulty by multiplying numerator and denominator by $\sqrt{2}$:

$$\frac{1}{\sqrt{2}} = \frac{\sqrt{2}}{2} = \frac{1 \cdot 414}{2} = \cdot 707.$$

We are sometimes instructed to "rationalize" such denominators as $\sqrt{3} - \sqrt{2}$ and $\sqrt{3} + \sqrt{2}$. Thus:

$$\frac{1}{\sqrt{3} - \sqrt{2}} = \frac{\sqrt{3} + \sqrt{2}}{(\sqrt{3} - \sqrt{2})\,(\sqrt{3} + \sqrt{2})} = \frac{1 \cdot 732 + 1 \cdot 414}{3 - 2}$$
$$= 3 \cdot 146.$$

But I must say that amongst many thousands of sums I have not yet come across one where the process was needed; apart of course from exercises devised to teach the process.

If many divisions by the same number have to be worked out it is convenient to do one division and then to work the rest as multiplication. Take a number at random, say 97. Suppose we often have to divide by 97. Division by 97 is the same as multiplication by $\frac{1}{97}$. Change $\frac{1}{97}$ to a decimal:

$$
\begin{array}{r}
\cdot010309 \\
97{\overline{)}\,1\cdot00} \\
97 \\
\hline
300 \\
291 \\
\hline
900 \\
873 \\
\hline
27 \\
\hline
\end{array}
$$

Hence to divide by 97 we can multiply by ·010309 or ·01031.

$3784 \div 97 = 37 \cdot 84$ (Move the decimal point two places
$1 \cdot 1352$ to the left.)
$\cdot03784$

$\overline{39 \cdot 01304 \text{ or } 39 \cdot 01}$

or $37 \cdot 84$
$1 \cdot 13$
$\cdot04$
$\overline{39 \cdot 01}$

The steps in the second process are:

(i) Move the decimal point two places to the left.

(ii) Cross off mentally the end figure (4); $8 \times 3 = 24$, carry 2; $7 \times 3 + 2 = 23$, write 3; $3 \times 3 + 2 = 11$, write 11.

(iii) Move the decimal point five places to the left. Discard all figures beyond the second decimal place; write 4 because the number is nearer 4 than 3.

If one often had to divide by 97 for any reason one could learn to apply the process quickly and automatically. Thus:

$$587643 \div 97 = 5876 \cdot 43$$
$$176 \cdot 29$$
$$5 \cdot 88$$
$$\overline{}$$
$$6058 \cdot 60 \text{ or } 6058$$

Bankers often have to divide by 73,000.

The number arises in finding the interest on a sum of money for a number of days. Suppose you have to find the interest on £89·4 for 45 days at 2 per cent. The interest is:

$$\frac{£89 \cdot 4 \times \cdot 02 \times 45}{365}$$

$$= \frac{£89 \cdot 4 \times 4 \times 45}{73,000}.$$

(We multiply numerator and denominator by 200. This changes ·02 to 4, and 365 to 73,000.)

4 is twice the rate per cent, and 45 is the number of days. So we have the rule: multiply the principal by twice the rate per cent and by the number of days; divide by 73 and 1000.

We want to find an easy way of dividing by 73.

```
                          ·01369863
                    73) 1·00
                        73
                       ───
                        270
                        219
                       ───
                        510
                        438
                       ───
                        720
                        657
                       ───
                        630
                        584
                       ───
                        460
                        438
                       ───
                        220
                        219
```

$\frac{1}{73}$ = ·01369863̇ or ·01369863 almost exactly.

$$\frac{1}{100} = ·01$$
$\frac{1}{3}$ of this = ·00333333
$\frac{1}{10}$ of this = ·00033333
$\frac{1}{10}$ of this = ·00003333

Add   ·01369999
Deduct $\frac{1}{10000}$ of the sum   ·00000136
                              ─────────
                              ·01369863

The result is $\frac{1}{73}$ almost exactly.

Hence to divide by 73: (i) Move the decimal point two places to the left. (ii) Add a third of this, a tenth of the second line, and again a tenth of the third line. (iii) For very accurate work deduct the ten-thousandth part of the answer.

This rule for dividing by 73 is an extremely neat example of a method that can often be used when a number of divisions have to be performed by the same number.

Multiplication does not always increase a number. If the multiplier is less than 1 we have a decrease. When we multiply by ·5 we get a half of the original number; when we multiply by ·1 we reduce the number to a tenth of itself; and so on. The square of a small number (less than 1) is a very small number; the cube is an extremely small number.

Thus: $\qquad ·001^2 = ·000001,$

$$·001^3 = ·000000001.$$

Even when a number is a very little less than one, continual multiplication of it by itself (raising it to higher and higher powers) finally reduces it to an exceedingly small quantity.

·99999 is a mere one-hundred-thousandth less than 1; if we multiply any number by it we deduct a one-hundred-thousandth of the number. Thus $·99999^2$ is a very little more than ·99998. $·99999^{10}$ is a little more than ·99990.

·99999 to the power 1000 is about ·99015.

·99999 to the power 1,000,000 is ·00005.

·99999 to the power 10,000,000 is the tenth power of ·00005.

We should have a string of 42 noughts before the significant figures begin. For the 100,000,000th power we should have over 400 noughts before the significant figures begin.

It does not matter how close to 1 a fraction may be—it may be as little as a quadrillionth less than 1—nevertheless continual multiplication finally reduces it to as near zero as we care to let it go.

We get an interesting case of multiplication when we square a number, like 1·001, which is a very little more than 1.

1·001	1·0012
1·001	1·0012
1·001	1·0012
·001001	·00120144
1·002001	1·00240144

Thus
$$1·001^2 = 1·002 \text{ (very nearly)},$$
$$1·003^2 = 1·006,$$
$$1·0012^2 = 1·0024.$$

We simply multiply the added little bit by 2, and ignore the very tiny bit that comes at the end.

If we use algebra we can get a general rule—that is one of the chief purposes of algebra:

$$(1+n)^2 = 1 + 2n + n^2.$$

If n is small n^2 is so small that we can ignore it, and say:

$$(1+n)^2 = 1 + 2n \text{ (if we can ignore } n^2).$$

Let us write the number and the algebraic expression together:

$$1·001^2 = 1·002001,$$
$$(1+n)^2 = 1 + 2n + n^2.$$

Now let us look at $1 \cdot 001^3 = 1 \cdot 001^2 \times 1 \cdot 001$:

$$1 \cdot 002001$$
$$1 \cdot 001$$
$$\overline{1 \cdot 002001}$$
$$\cdot 001002001$$
$$\overline{1 \cdot 003003001}$$

Thus: $1 \cdot 001^3 \ = 1 \cdot 003$ (very nearly),

$1 \cdot 004^3 \ = 1 \cdot 012$,

$1 \cdot 0012^3 = 1 \cdot 0036$.

We multiply the little bit at the end by 3, and ignore what comes after.

Let us compare $1 \cdot 001^3$ with $(1+n)^3$:

$$1 \cdot 001^3 = 1 \cdot 003003001,$$
$$(1+n)^3 = 1 + 3n + 3n^2 + n^3.$$

There is no need to multiply out $1 \cdot 001^4$ or $1 \cdot 001^5$. We can write them down by comparison with $(1+n)^4$ and $(1+n)^5$:

$$(1+n)^4 = 1 + 4n + 6n^2 + 4n^3 + n^4,$$
and $1 \cdot 001^4 = 1 \cdot 004006004001;$

$$(1+n)^5 = 1 + 5n + 10n^2 + 10n^3 + 5n^4 + n^5,$$
and $1 \cdot 001^5 = 1 \cdot 005010010005001.$

So we can write: $1 \cdot 001^2 = 1 \cdot 002,$

$1 \cdot 001^3 = 1 \cdot 003,$

$1 \cdot 001^4 = 1 \cdot 004,$ and so on.

In every case we multiply the little bit at the end by the index of the power (2, 3, 4, or whatever it is). And we have

to be sure that we really can neglect the square and higher powers.
$$1 \cdot 1^2 = 1 \cdot 21.$$

To write $1 \cdot 1^2 = 1 \cdot 2$, we should have to be sure that we can neglect $\cdot 01$.
$$1 \cdot 9^2 = 3 \cdot 61.$$
$$1 + 2 \times \cdot 9 = 2 \cdot 8.$$

It is quite obvious that we cannot neglect $\cdot 9^2 = \cdot 81$.

The sort of approximation we have been considering is useful when we are dealing with such small quantities as the expansion of solids by heating. The expansion is a small fraction of the length at $0°$ C. To find the increase in area we should have to square the length. Thus a length of 1 unit might increase to $1 \cdot 0001$ units. The area would increase from 1 square unit to $1 \cdot 0001^2$ square units.

$1 \cdot 0001^2 = 1 \cdot 0002$, so that the coefficient of areal expansion is twice as great as the coefficient of linear expansion.

$1 \cdot 0001^3 = 1 \cdot 0003$, so that the coefficient of volume expansion is three times the coefficient of linear expansion.

I have just looked up the coefficient of expansion of copper in the *Encyclopaedia Britannica*. The latest edition gives it as $0 \cdot 001869$ per degree centigrade (roughly $0 \cdot 002$). That seems rather a lot. If we imagine our copper kettle being heated from freezing point up to boiling point the increase in lengths would be a hundred times as much, that is $\cdot 2$. Now we all know that our copper kettles do not add a fifth to their stature when we boil them. So I turned to a reference book of chemical and physical constants; and I found the coefficient of expansion of copper to be $\cdot 0000167$—which certainly seems more probable.

Now the coefficient of expansion is used in this way: suppose we start with a standard length—the length of a rod at $0°$ C. For every degree increase of temperature the

rod increases in length by ·0000167 of its length at 0° C.
To take an easy example, let us start with a rod 1 metre long
at 0° C. At 1° C. its length would be 1·0000167 metres; at
2° C. the length would be 1·0000334 metres; at 100° C. it
would be 1·00167 metres; and so on.

We might want to find the increase in area of a copper
object when it is heated. The coefficient of areal expansion
is ·0000167 × 2 = ·0000334. Thus an area of 1 square
metre at 0° C. would increase to 1·00334 at 100° C. More
often we need to use the coefficient of volume expansion.
This is three times ·0000167 = ·0000501. A volume of 1 cubic
metre at 0° C. increases to 1·00501 cubic metres at 100° C.

Let us look at that *Encyclopaedia* kettle again. Its coefficient
of volume expansion is 3 × ·001869 = ·005607. For 100
degrees increase in temperature the increase in volume is
·5607; so that the *Encyclopaedia* kettle holds more than half
as much again when it boils. Actually the increase in
volume would be more than that, because we have got above
the limit at which we can neglect the square and cube of the
increase in length. 1 unit of length increases to 1·1869 units
for 100 degrees rise of temperature; and so the volume
increases from 1^3 to $1·1869^3 = 1·67$. Our *Encyclopaedia* kettle
holds two-thirds as much again.

Lengths in an ordinary kettle would increase from 1 to
1·00167 for 100 degrees increase in temperature, and the
volume from 1 to 1·00501. That would be an increase of
5 pints in 1000 pints, as compared with 670 pints in 1000
pints in the *Encyclopaedia* kettle.

The error in giving the coefficient seems so trivial that
apparently it was not noticed. And we have seen what it
jumps to when we begin to multiply. It is one of those
trivial, unnoticed, errors that can upset a whole Utopia.

XI

Tables

THE Infants Schools and Kindergartens—which usually do things well—start little children constructing tables. Curiously, this admirable practice is dropped in the higher schools, and one of the most important parts of arithmetic is almost entirely neglected. The construction of tables is one of the most important contributions of arithmetic to commerce.

Children are taught how to find the value of, say, 1 ton 14 cwt. 30 lb. at £2. 15*s*. per ton. It is a bare possibility that such a sum might be needed—perhaps in finding the cost of a load of coal. But it would be an absurd waste of energy to work out the cost of each load by the school method. What is wanted is the ability to construct a table:

	£	s.	d.			s.	d.
1 ton	2	15	0		110 lb.	2	8¼
10 cwt.	1	7	6		100	2	5½
9		1	4	9	90	2	2½
8		1	2	0	80	1	11½
7			19	3	70	1	8½
6			16	6	60	1	5½
5			13	9	50	1	2¾
4			11	0	40		11¾
3			8	3	30		8¾
2			5	6	20		5¾
1			2	9	10		2¾
					1		¼

The cost of 1 ton 14 cwt. 30 lb.

$$= £2. \ 15s. \ 0d. + £1. \ 7s. \ 6d. + 11s. \ 0d. + 8\tfrac{3}{4}d.$$
$$= £4. \ 14s. \ 2\tfrac{3}{4}d.$$

And we can call that either £4. 14s. 2d. or £4. 14s. 3d. The construction of the table is easy enough:

$$\text{Cost of 1 cwt.} = £2. \ 15s. \div 20$$
$$= 2s. \ 9d.$$

Enter 2s. 9d. beside 1 cwt. and continually add 2s. 9d. for 2 cwt., 3 cwt., etc. There is a check at 10 cwt., since the cost of 10 cwt. $= £2. \ 15s. \div 2$.

$$\text{Cost of 1 lb.} = 2s. \ 9d. \div 112$$
$$= \cdot 295d.$$

Multiply ·295d. by 10, 20, etc. and enter the result to the nearest farthing below.

Here is a tentative list of tables that might be constructed:

Multiplication tables up to 20 times.

Tables of squares from 1·0, proceeding by tenths up to 10·0.

A similar table of cubes.

Factors of all numbers up to 200.

Divisors of numbers up to 200.

Square and cube roots of numbers up to 100.

Trigonometrical ratios.

Logarithms of numbers from 1·0, proceeding by tenths up to 10·0.

Ready reckoners for various prices: 1 to 10, then by 10's to 100, then by 100's to 1000.

Interest on £1, £2, etc. up to £10; then £20, £30, etc. up to £100, at various rates per cent per annum.

Discounts at various percentages or amounts in the pound, on 5*s*., 10*s*., 15*s*., and then by pounds up to £10, and by 10's up to £100.

Wages tables at various rates: ¼ hr., ½ hr., ¾ hr., 1 hr., 2 hr., etc. to 10 hr.; then 20 hr., 30 hr., 40 hr., 50 hr.

Areas of circles of radii 0·1 in., 0·2 in., etc. proceeding by tenths up to 10·0 in.

A similar table for the volumes of spheres.

The list could be extended, almost indefinitely, to cover all cases where many calculations of the same kind have to be made.

As an example of the method of constructing a rather elaborate table we will consider the construction of an interest table.

We want to be able to find from the table the interest on any number of pounds for any number of days up to 365. It may seem as though we should have to have a separate table for each number of days, but there is a means of overcoming this difficulty. The interest on £100 for 1 day is the same as the interest on £1 for 100 days. In the table we need only give the interest for 100. The 100 may stand for £100 for 1 day, £50 for 2 days, £25 for 4 days, and so on.

We should, however, need a separate table for each rate per cent. We will take here interest at 2 per cent. We begin by making an exact calculation of the interest on £100 for 1 day. Exactness is essential because we shall have to multiply this interest by large numbers. We use the "third, tenth and tenth" rule.

$$\text{Interest} = \frac{\text{principal} \times \text{twice rate per cent} \times \text{number of days}}{73,000}$$

Interest on £100 for 1 day at 2 per cent

$$=\frac{£100 \times 4 \times 1}{73,000} = \frac{£\cdot4}{73}.$$

£·004	($\frac{1}{100}$)
·001333333	($\frac{1}{3}$ of this)
·000133333	($\frac{1}{10}$ of this)
·000013333	($\frac{1}{10}$ of this)
·005480000	(since the 9's run on)
·000000548	(deduct $\frac{1}{10000}$th)
·005479452	

The interest on 100 is £·005479452. We put this amount at the top of the table, and we continually add it in order to get the interest for 200, 300, and so on. The table as far as 1000 is:

	£	s.	d.
100	·005479452		$1\frac{1}{4}$
200	·010958904		$2\frac{1}{2}$
300	·016438356		$3\frac{3}{4}$
400	·021917808		$5\frac{1}{4}$
500	·027397260		$6\frac{1}{2}$
600	·032876712		$7\frac{3}{4}$
700	·038356164		9
800	·043835616		$10\frac{1}{2}$
900	·049315068		$11\frac{3}{4}$
1000	·054794520	1	1

The third column contains the interest to the nearest farthing below. At 1000 there is a check on the working since this line is 10 times the top line. We now proceed in thousands—2000, 3000, etc. up to 100,000. There is a check at 10,000 (£·5479452); another check at 20,000 (= 2000 × 10, hence £1·0958904); and so on at each 10,000. Beyond 100,000 the table proceeds by hundreds of thousands to 1,000,000.

When the table is printed for use, only the first and third columns are given. The hundreds, being of lesser importance, are placed in the lower right-hand corner of the page.

In using this table we multiply the number of pounds by the number of days and find the product in the table. Thus to find the interest on £80 for 17 days at 2 per cent we have:

$$80 \times 17 = 1360 \quad \text{(Ignore the 60.)}$$

Interest on 1000 = 1s. 1d.
Interest on   300 =   3¾d.
                    ‾‾‾‾‾‾‾‾
                    1s. 4¾d.   (Ignore the ¾d.)

There is of course no need to work out each item in an account separately. The products are added, and one calculation of interest from the table suffices.

A table giving the number of days from each day of the year to the end of the year is extremely useful in calculations of interest. In finding the number of days from one date to another it is only necessary to subtract the number of the second date from the number of the first. The table is:

Jan. 1	364	Feb. 1	333	Mar. 1	305	etc.
2	363	2	332	2	304	
3	362	3	331	3	303	
etc.		etc.		etc.		

Interest tables are also used which give the interest on £1, £2, £3, etc. for any number of days. The amounts of money go in single pounds from £1 to £100, and then in hundreds to £1000. In a book of tables there would be pages for 1 day, 2 days, 3 days, etc. up to 365 days.

These tables are constructed in a similar way to the previous table.

5 *per cent tables*

Principal £		Interest for 20 days £ s. d.		
1	·002739726			½
2	·005479452			1¼
3	·008219178			1¾
4	·010958904			2½
5	·013698630			3¼
6	·016438356			3¾
7	·019178082			4½
8	·021917808			5¼
9	·024657534			5¾
10 etc. to	·027397260			6½
100 etc. to	·2739726		5	5¾
1000	2·739726	2	14	9½

Only the first and third columns are given in the printed tables. The middle column is used in constructing the tables.

We will now explain the construction of a table for turning angles given as degrees into radians.

$$180° = 3·14159265 \text{ radians.}$$
By division $\quad 1° = ·0174533$ radian.

This is a close enough approximation to give the results to five figures. The method of construction is by continually adding ·0174533 to the previous result.

		radians
1°	·0174533	·01745
2°	·0349066	·03491
3°	·0523599	·05236
4°	·0698132	·06981
5°	·0872665	·08727
etc.		

The first and third columns only are given in the table. There should also be entered somewhere on the page the values of minutes in radians:

1'	·00029	10'	·00291	35'	·01018
2'	·00058	15'	·00436	40'	·01164
3'	·00087	20'	·00582	45'	·01309
4'	·00116	25'	·00727	50'	·01454
5'	·00145	30'	·00873	55'	·01600

A more elaborate table is sometimes used which gives the value in radians for each minute:

'	0°	1°	2° etc.		
0	·0000000	·0174533	·0349066	0	·0000000
1	·0002909	·0177442	·0351975	1	·0000048
2	·0005818	·0180351	·0354884	2	·0000097
3	·0008727	·0183260	·0357792	3	·0000145

etc. to 60'.

To change 2° 3' 1'' into radians we run a finger down the column headed 2° to the line opposite 3' (i.e. ·0357792) and we add the value of 1'' in radians:

$$·0357792$$
$$·0000048$$
$$2°\ 3'\ 1'' = ·0357840 \text{ radian}$$

A table of sines of angles proceeding by degrees from 0° to 45° may be calculated from the series:

$$\sin\theta = \theta - \frac{\theta^3}{\underline{3}} + \frac{\theta^5}{\underline{5}} - \frac{\theta^7}{\underline{7}} + \cdots$$

θ is the value of the angle in radians. $\underline{3}$ (factorial 3) stands for $1 \times 2 \times 3$; $\underline{5}$ stands for $1 \times 2 \times 3 \times 4 \times 5$; and so on.

Thus $\qquad$ sin $1° =$ sin $\cdot$0174533 radian

$$= \cdot0174533 - \frac{\cdot0174533^3}{1 \times 2 \times 3} + \frac{\cdot0174533^5}{1 \times 2 \times 3 \times 4 \times 5} - \ldots$$

Actually for the sine of $1°$ we need only the first term to give the result true to five figures. Nevertheless the terms should be worked out with considerable accuracy—for the following reason:

$$\text{sin } 2° = \text{sin } \cdot0349066 \text{ radian}$$

$$= \text{sin } \cdot0174533 \times 2 \text{ radians}$$

$$= \cdot0174533 \times 2 - \frac{(\cdot0174533 \times 2)^3}{1 \times 2 \times 3} + \frac{(\cdot0174533 \times 2)^5}{1 \times 2 \times 3 \times 4 \times 5} - \ldots$$

We can get the second term by multiplying the calculated value of $\frac{\cdot0174533^3}{1 \times 2 \times 3}$ by 2^3 ($=8$), the third term by 2^5 ($=32$), and so on. For $3°$ we should multiply $\frac{\cdot0174533^3}{1 \times 2 \times 3}$ by 3^3 for the second term, and so on again.

The table we have been considering is not too easy to construct. Fortunately the labour can be lightened by constructing a table of cosines at the same time. We use the series:

$$\cos \theta = 1 - \frac{\theta^2}{\lfloor 2} + \frac{\theta^4}{\lfloor 4} - \frac{\theta^6}{\lfloor 6} + \ldots$$

$$\cos 1° = \cos \cdot0174533 \text{ radian}$$

$$= 1 - \frac{\cdot0174533^2}{\lfloor 2} + \text{etc.}$$

$$= 1 - \cdot0001523 + \text{negligible terms}$$

$$= \cdot9998477.$$

Or we can use: $\qquad \cos 1° = \sqrt{1 - \sin^2 1°}$.

Now we know from trigonometry that:

$$\sin (A+B) = \sin A . \cos B + \cos A . \sin B$$
$$\sin 2° = \sin (1° + 1°)$$
$$= \sin 1° \cos 1° + \cos 1° \sin 1°$$
$$= 2 \sin 1° \cos 1°$$
$$= 2 \times ·0174524 \times ·9998477$$
$$= ·034900.$$

Also
$$\cos (A+B) = \cos A . \cos B - \sin A . \sin B$$
$$\cos 2° = \cos (1° + 1°)$$
$$= \cos^2 1° - \sin^2 1°$$
$$= (\cos 1° + \sin 1°)(\cos 1° - \sin 1°)$$
$$= 1·0173001 \times ·9823953$$
$$= ·9993908.$$

We now proceed to find sin 3° and cos 3°:

$$\sin 3° = \sin 2° \cos 1° + \cos 2° \sin 1°,$$

and as we know the sines and cosines of 1° and 2° we can find the value of sin 3°:

$$\cos 3° = \cos 2° \cos 1° - \sin 2° \sin 1°,$$

and we can calculate this also.

And so we can proceed step by step through the table. We have already seen that we need not go further than 45°, since

$$\sin 46° = \cos (90° - 46°)$$
$$= \cos 44°,$$

and equally $\cos 46° = \sin 44°$, and so on.

After a table of this kind most of the tables required in commerce are mere child's play. Many of them merely require the continued addition of sums of money, as in this wages table:

Hrs.	1s. £	s.	d.	1s. 1d. £	s.	d.	1s. 2d. £	s.	d.	1s. 3d. £	s.	d.
$\frac{1}{4}$			3			$3\frac{1}{4}$			$3\frac{1}{2}$			$3\frac{3}{4}$
$\frac{1}{2}$			6			$6\frac{1}{2}$			7			$7\frac{1}{2}$
$\frac{3}{4}$			9			$9\frac{3}{4}$			$10\frac{1}{2}$			$11\frac{1}{4}$
1		1	0		1	1		1	2		1	3
2		2	0		2	2		2	4		2	6
etc.												

XII

Units

WHEN we are measuring, we want a convenient amount of the quantity which is being measured, as a unit for measuring. We have already seen how this idea of a convenient quantity runs all through the English system of weights and measures. There are people who imagine the English weights and measures to be haphazard and devoid of system; but they are usually the intelligentsia, who have not even considered what a unit is—the kind of people who swallow "intelligence tests" with their mouths open. A rigid decimal system has its advantages; for scientific purposes, it enables the exact degree of approximation to be stated; and presumably such a system is easier for peoples of inferior arithmetical ability. But for convenience in everyday working there is no comparison between the English weights and measures, and the inflexible decimal system. If the British people were to give up their weights and measures, for a mess of decimal pottage, they would be abandoning a wonderful system, devised by their Anglo-Saxon forbears who were arithmetical geniuses.

People who try to persuade us to abandon our *ad hoc* system might reflect that we are always trying to devise suitable *ad hoc* units. We choose a unit to fit a particular kind of measurement, quite irrespective of whether it fits in or not with the metric system.

When numbers get inconveniently large or inconveniently small it is a very usual practice to find a more suitable unit, one which will reduce the numbers to a reasonable size. We choose a large unit for large quantities, and a small unit for small quantities; and it is a matter of indifference whether the unit fits in with the metric system or not; sometimes it does and sometimes it doesn't.

When we have a number of large measurements it is a common practice to use one of them as the unit. We generally use as a unit a quantity of which we have some knowledge, or in which we have some special interest. Most people have some vague idea of the size of England, so that we can use either the length of England or the area of England as a unit.

The area of England and Wales is 58,000 square miles. The area of India is 1,805,000 square miles. Using the area of England and Wales as unit, the area of India is:

$$\frac{1,805,000}{58,000} = \text{about } 31.$$

We can say if we like that India is 31 times as great as England and Wales; or we can give the area as 31 with the area of England and Wales as unit. The area of Canada is 3,690,000 square miles, so that the area with the unit of 58,000 square miles is:

$$\frac{3,690,000}{58,000} = \text{about } 63.$$

Australia has an area of 2,975,000 square miles, and

$$\frac{2,975,000}{58,000} = \text{about } 51.$$

New Zealand $\qquad \dfrac{105,000}{58,000} = 1 \cdot 8.$

Union of South Africa $\dfrac{472,000}{58,000} = 8.$

So we have the comparison:

England and Wales	1	India	31
Canada	63	Australia	51
New Zealand	1·8	South Africa	8

For measurements in the solar system we often use earth measurements as units. The following list gives the distances of the planets from the sun, (*a*) in millions of miles, (*b*) with the earth's distance from the sun as unit.

	Mercury	Venus	Earth	Mars
(*a*)	36·0	67·2	92·9	141·6
(*b*)	·39	·72	1	1·52

	Jupiter	Saturn	Uranus	Neptune
(*a*)	483·3	886·2	1782·8	2793·5
(*b*)	5·20	9·54	19·19	30·07

The numbers in the second lines are obtained by dividing the numbers in the upper lines by 92·9. In some tables the distance for the earth would be given as 1·00. But this seems to me to disguise the fact that the earth's distance was chosen as a unit. 1·00 makes it appear as if it just happened to be that to the nearest hundredth.

The use of such a large unit makes the numbers more comprehensible, and it does draw attention to the fact that Neptune is thirty times as far off from the sun as the earth.

The choice of a very large unit is especially useful in a case like the masses of the sun and planets where the measurements run to trillions of tons. We know the following facts about the sun and planets:

	Radius in miles	Radius Earth = 1	Density Earth = 1
Sun	433,000	109·2	·25
Mercury	1,387	·35	·88
Venus	3,783	·955	·94
Earth	3,963	1	1
Mars	2,110	·53	·71
Jupiter	43,850	11·1	·25
Saturn	38,170	9·63	·12
Uranus	15,440	3·90	·24
Neptune	16,470	4·15	·23

If we want to find the actual densities we have to multiply the numbers in the last column by the density of the earth ($5\frac{1}{2} \times 62\frac{1}{4}$ lb. per cubic foot). Notice how the table as it stands draws attention to the similarity in density of the four innermost planets, and the comparative lightness of the four outermost planets.

We can use the table to compare the masses of the planets without going to the trouble of calculating them:

$$\text{mass} = \text{volume} \times \text{density},$$

and since the volume is proportional to the cube of the radius:

$$\text{mass} = kr^3 \times \text{density}.$$

In the case of the earth we have made $r = 1$ and $d = 1$.

$$\therefore \quad \text{Mass of earth} = k.$$

If we call the earth's mass 1, then $k = 1$.

And so, using the earth's mass as unit, the mass of the sun or any planet is given by:

$$\text{mass} = r^3 d.$$

r is the radius (earth = 1) and d is the density (earth = 1). Thus for the sun we have $r = 109\cdot2$ and $d = \cdot25$:

$$
\begin{array}{r}
109\cdot2 \\
109\cdot2 \\
\hline
10920\cdot0 \\
982\cdot8 \\
21\cdot8 \\
\hline
11924\cdot6 \\
109\cdot2 \\
\hline
1192460\cdot0 \\
107321\cdot4 \\
2384\cdot9 \\
\hline
4)\,1302166\cdot3 \\
\hline
325541\cdot6
\end{array}
$$

The mass of the sun is nearly 326,000 times as great as that of the earth.

The masses of the planets, on the same scale, are:

Mercury	Venus	Earth	Mars	Jupiter	Saturn	Uranus	Neptune
·34	·82	1	·11	314	94	14·4	16·7

The combined mass of all the planets is the sum of these numbers:

$$= 441\cdot4.$$

So that the whole of the planets have a combined mass about 440 times that of the earth. The mass of Jupiter is $\dfrac{314}{440} = \cdot71$ of the whole, not far short of $\frac{3}{4}$ of the whole. Also the mass of the sun is $\dfrac{325540}{441\cdot4} =$ about 740 times as great as the masses of all the planets. They are a mere 740th of the mass of the sun.

As the quantities we are measuring increase in size we often try to find a larger unit to measure them with. We could of course give any distance at all in, say, tenths of an inch. The distance of the nearest star is:

$$26 \text{ billion miles} = 26 \times 10^{12} \times 5280 \times 12 \times 10 \text{ tenths of an inch}$$
$$= 2 \cdot 6 \times 5 \cdot 28 \times 1 \cdot 2 \times 10^{18} \text{ tenths of an inch}$$
$$= 16 \cdot 5 \times 10^{18} \text{ tenths of an inch}$$
$$= 16\tfrac{1}{2} \text{ trillion tenths of an inch.}$$

But that does not seem very helpful as an attempt at a picture of the distance. It was to give some sort of a picture of the distance that the light-year and the parsec were devised.

We have to keep in mind that measurements have two distinct purposes—they may be mere records (in which case the particular unit of measurement does not matter very much); or they may be attempts to picture the quantities measured (in which case the choice of a suitable unit is as important as the measurement itself). We can, and do, give all lengths, however great or small, in centimetres. But we find other units when the purpose is to provide some sort of a picture of measurements.

The confusion of these two ideas of units leads to all sorts of difficulties and oddities. Nearly all the difficulties that arise in school attempts to interpret our *ad hoc* system of weights and measures arise from this confusion. For recording purposes we never need more than two units, and we can always do with one only. The furlong and pole are descriptive units, of the same kind as the light-year. To include them in a long string of intermediate units serves no purpose at all; it is not necessary as a record, and it is absurd as a description. We might just as well give the distance of a star as so many light-years, so many semi-major axes of the earth's orbit, so many equatorial semi-diameters of the earth —all good descriptive units, but absurd when put into a jumble.

We have been considering extremely great units, suitable for the measurement of extremely great lengths. At the other end of the scale we need very small units for measuring such small quantities as the wave-lengths of light. There are three of these extremely small units in common use. The largest of them, or rather the least small, is the *micron*. This is an extremely small unit—the millionth part of a metre— 25,000 of them go to the inch. The micron is sometimes written as μ; thus 25·3 microns would be written 25·3μ. A still smaller unit is $\mu\mu$ which stands for the thousand-millionth part of a metre—the thousandth part of a micron. A still smaller unit is the Ångström unit (Å.U.) which is the ten-thousand-millionth part of a metre.

$$\text{The micron} = 10^{-6} \text{ metre,}$$
$$\mu\mu = 10^{-9} \text{ metre,}$$
$$\text{Å.U.} = 10^{-10} \text{ metre.}$$

Because its value is 10^{-10} metre the Ångström unit is sometimes called a tenth-metre. This is not to be confused with a tenth of a metre:

$$\text{tenth of a metre} = 0\cdot1 \text{ m.,}$$
$$\text{tenth-metre} = 0\cdot0000000001 \text{ m.}$$

The wave-lengths of red light range from about 7700 tenth-metres to 6500 tenth-metres; the ultra-violet starts at about 3600 tenth-metres.

There is no great difficulty in changing from one unit to

another. Thus we might change 7700 tenth-metres to metres
by moving the decimal point ten places to the left:

$$7700 \text{ tenth-metres} = \cdot00000077 \text{ m.}$$

$$= \cdot000077 \text{ cm.}$$

We might want to change miles per hour to feet per
second:

$$60 \text{ miles per hour} = 60 \times 5280 \text{ ft. per hour}$$

$$= \frac{60 \times 5280}{3600} \text{ ft. per sec.}$$

$$= 88 \text{ ft. per sec.}$$

In all such cases a factor may be found that will perform
the required change:

$$1 \text{ mile per hour} = \tfrac{5280}{3600} \text{ ft. per sec.}$$

$$= \tfrac{22}{15} \text{ ft. per sec.}$$

Hence to change miles per hour to feet per second we
multiply the number of miles per hour by $\tfrac{22}{15}$. To change
feet per second to miles per hour we multiply by $\tfrac{15}{22}$. Thus:

$$88 \text{ feet per second} = 88 \times \tfrac{15}{22} \text{ miles per hour}$$

$$= 60 \text{ miles per hour.}$$

Suppose we want a factor to change pounds per ton to
pence per lb.

$$£1 \text{ per ton} = \tfrac{240}{2240}d. \text{ per lb.}$$

$$= \tfrac{3}{28}d. \text{ per lb.}$$

Hence to change pounds per ton to pence per lb. we multiply
by $\tfrac{3}{28}$:

$$£7 \text{ per ton} = 7 \times \tfrac{3}{28}d. \text{ per lb.}$$

$$= \tfrac{3}{4}d. \text{ per lb.}$$

To change pence per lb. to pounds per ton we multiply by $\frac{28}{3}$.

$$9d. \text{ per lb.} = £9 \times \tfrac{28}{3} \text{ per ton}$$
$$= £84 \text{ per ton.}$$

Let us look at a more difficult example. Suppose we want a factor to change shillings per yard to francs per metre. Francs are quoted at so many to the pound, say $£1 = f$ francs. 1 metre $= 39\cdot37$ inches:

$$1s. \text{ per yard} = £\tfrac{1}{20} \text{ per yard}$$

$$= \frac{f}{20} \text{ francs per yard}$$

$$= \frac{f}{20} \times \frac{39\cdot37}{36} \text{ francs per metre}$$

$$= \cdot05468 f \text{ francs per metre}$$

$$xs. \text{ per yard} = \cdot05468xf \text{ francs per metre.}$$

Hence to change any number of shillings per yard to francs per metre: multiply the number of shillings by the number of francs to the pound and then by $\cdot05468$. For this purpose it would be useful to have a table of multiples of $\cdot05468$:

$1s.$	$\cdot05468$
$2s.$	$\cdot10936$, and so on to $20s.$
$1d.$	$\cdot00456$
$2d.$	$\cdot00911$, and so on to $11d.$

It is then only necessary to multiply the number in the table (or shillings + pence) by the number of francs to the pound.

XIII

Oddities of Numbers

THERE are parts of arithmetic that deal with the oddities of numbers. They are seldom of any practical importance; one does them for the fun of the thing, or not at all. They are of the same order of things as poetry that is amusing and fanciful without going too deep.

I have always been glad of recurring decimals, because they brought the first touch of romance into arithmetic for me. You subtracted ·9̇ from 1 and you got ·00000... with an elusive 1 that must come somewhere but that did not seem to come anywhere.

There are two kinds of oddities in numbers. Some are essential properties of the numbers. 16 objects can always be arranged as 4 fours, no matter how the number is expressed. We can write 16 as a dozen and 4 (14, where the 1 stands for a dozen and not for 10) or as 2 eights (20, where the 2 stands for 2 eights); and so on. But however it is expressed $16 = 4 \times 4$.

A great many properties of numbers depend on the scale in which the number is expressed. Take a number at random: 873426. $8+7+3+4+2+6=30$; $3+0=3$. We know without further inquiry that the remainder when 873426 is divided by 9 is 3. That method and the result depend on the scale in which the number is expressed. If the number were written in powers of 12 instead of in powers of 10 the same method would apply to division by eleven.

To turn $\frac{1}{7}$ into a decimal we divide 1 by 7:

$$326451$$
$$7\overline{)\,1\cdot000000}$$
$$\cdot 142857\ldots$$

The remainders have been placed over the top. There are only six possible remainders; when they are all exhausted we start again with 1, and so the series of numbers repeats.

If we start with 3 instead of 1 we get the same series, but starting with 4, that is ·428571.... Hence it arises that, if we multiply 142,857 by any number up to 6, we get the same series of figures:

$$142,857 \times 4 = 571,428, \quad \text{and so on.}$$
$$142,857 \times 7 = 999,999.$$

Hence $\cdot 142857 \times 7 = \cdot \dot{9}.$

If we multiply 142,857 by any number from 8 to 99 we usually get the same series of figures, except that one of the digits in the series is replaced by two figures whose sum is equal to this digit:

$$142,857 \times 19 = 2,714,283.$$

Starting with 1 we have 1, 4, 2, 8, 3+2=5, 7. There are exceptions, for example, when one of the digits in the multiplier is 7, or when the outside numbers add up to more than 9: $142,857 \times 37 = 5,285,709.$

Add the outside digits 5+9=14; write 14 instead of 5 and we again have 142,857.

Apart from 10's and 5's the only denominators that give terminating decimals are powers of 2:

$$\tfrac{1}{2} = \cdot 5; \quad \tfrac{1}{4} = \cdot 25; \quad \tfrac{1}{8} = \cdot 125; \quad \text{and so on.}$$

This is of course because 2 and 5 are the only factors of 10.

When we divide a number by 13 there are twelve possible remainders. Six of these remainders occur in $1 \div 13$ and the other six in $2 \div 13$. $1 \div 13 = \cdot 076923$. The remainders are 10, 9, 12, 3, 4, 1. $2 \div 13 = \cdot 153846$. The remainders are 7, 5, 11, 6, 8, 2. The sum of the remainders in each case is 39. No remainder could belong to both sets; if it did then the other set of numbers would begin to recur.

Thus we get two sets of digits for thirteenths: 076923 for $\frac{1}{13}$, $\frac{3}{13}$, $\frac{4}{13}$, $\frac{9}{13}$, $\frac{10}{13}$, $\frac{12}{13}$ (the numbers, of course, of the remainders), and 153846 for $\frac{2}{13}$, $\frac{5}{13}$, $\frac{6}{13}$, $\frac{7}{13}$, $\frac{8}{13}$, $\frac{11}{13}$. Multiply the first set by 2, 5, 6, 7, 8, or 11 and we get the second set. Multiply the second set by $1\frac{1}{2}$, 2, $4\frac{1}{2}$, 5 or 6, and we get the first set. (The halves come in because $\cdot i 5384\dot{6} = \frac{2}{13}$ and not $\frac{1}{13}$.) Multiply the first set by 13 and we get a string of 9's.

$\cdot \dot{1}$ means $\cdot 11111 1 \ldots$
$$= \tfrac{1}{10} + \tfrac{1}{100} + \tfrac{1}{1000} + \ldots.$$

That is a geometrical series; each fraction is obtained from the previous one by multiplying it by a fixed number ($\frac{1}{10}$). It is quite easy to find the sum of this series, even though it runs on indefinitely, because the terms rapidly get less.

Let us call the sum of the series S:
$$S = \tfrac{1}{10} + \tfrac{1}{100} + \tfrac{1}{1000} + \ldots$$
and
$$\tfrac{1}{10}S = \qquad \tfrac{1}{100} + \tfrac{1}{1000} + \ldots.$$

(We have only to multiply term by term by $\frac{1}{10}$.)

Take the second line from the first:

$\frac{9}{10}S = \frac{1}{10}$. (The elusive term at the end is indefinitely small—$\cdot \dot{0}$.)
$$9S = 1, \text{ or } S = \tfrac{1}{9}.$$

So that $\cdot \dot{1} = \frac{1}{9}$. We can check the result by finding:

$$1 \div 9 = \cdot 1111 \ldots.$$

Any other recurring decimal can be dealt with in a similar way. Take $\cdot\dot{3}\dot{7}$ for example:

$$\cdot\dot{3}\dot{7} = \cdot373737\cdots$$
$$= \tfrac{37}{100} + \tfrac{37}{10000} + \tfrac{37}{1000000} + \cdots$$
$$= 37\left(\tfrac{1}{100} + \tfrac{1}{10000} + \tfrac{1}{1000000} + \cdots\right).$$

The fixed number here is $\tfrac{1}{100}$.

$$S = \tfrac{1}{100} + \tfrac{1}{10000} + \tfrac{1}{1000000} + \cdots,$$
$$\tfrac{1}{100}S = \qquad\ \ \tfrac{1}{10000} + \tfrac{1}{1000000} + \cdots,$$
$$\tfrac{99}{100}S = \tfrac{1}{100},$$
or $\qquad\qquad S = \tfrac{1}{99}.$

Hence $\cdot\dot{3}\dot{7} = 37S = \tfrac{37}{99}$; and again $37 \div 99 = \cdot3737\cdots,$

$$\cdot\dot{0}\dot{9} = \tfrac{9}{99} = \tfrac{1}{11}; \text{ and } 1 \div 11 = \cdot0909\ldots.$$

Suppose we have a decimal only part of which recurs, $\cdot25\dot{4}\dot{9}$ for example.

$\cdot25\dot{4}\dot{9}$ means of course $\cdot25494949\ldots.$

We can find its value in this way. Let us call the value x:

$$x = \cdot25\dot{4}\dot{9}.$$

We want to get rid of the $\cdot25$ so as to have a pure recurring decimal. We need only multiply by 100:

$$100x = 25\cdot4949\ldots \quad \text{or } 25\cdot\dot{4}\dot{9}$$
$$= 25\tfrac{49}{99}$$
$$x = \tfrac{25}{100} + \tfrac{49}{9900}$$
$$= \frac{25 \times 99 + 49}{9900}.$$

Here is another little device:

$$= \frac{25 \times 100 - 25 + 49}{9900}$$
$$= \frac{2549 - 25}{9900} = \frac{2524}{9900}.$$

That gives the rule which used to be employed to worry schoolchildren: write down the whole of the decimal (2549); subtract the part which does not repeat—that gives the numerator (2549−25=2524). For the denominator write a 9 for each figure that recurs followed by 0 for each non-recurring figure. But the interest is in the method; the rule has no practical importance, and no one need remember it.

Let us write the squares of the numbers 0, 1, 2, 3, etc. We have:

$$0, \quad 1, \quad 4, \quad 9, \quad 16, \quad 25, \quad 36, \quad \text{etc.}$$

Now subtract each number from the one following it. This gives:

$$1, \quad 3, \quad 5, \quad 7, \quad 9, \quad 11, \quad \text{etc.}$$

and if we do a second series of subtractions we get:

$$2, \quad 2, \quad 2, \quad 2, \quad 2, \quad \text{etc.}$$

The drawing shows how the series of odd numbers arises.

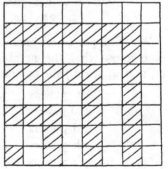

To change 1^2 to 2^2 we add 1 to each side and the odd 1 in the corner; that is $1+(1+1)$ or $1+2=3$. To change 2^2 to 3^2 we add $2+(2+1)$ or $2+3=5$; to change 3^2 to 4^2 we add $3+(3+1)$ or $3+4=7$; and so on.

To change 21^2 to 22^2 we add $21+22$:

$$21^2=441,$$
$$22^2=441+21+22$$
$$=484.$$
$$484+22+23=529$$
$$=23^2 \quad \text{and so on.}$$

Now look at the series of cubes:

$$0, \quad 1, \quad 8, \quad 27, \quad 64, \quad 125, \quad 216, \quad 343, \quad \text{etc.}$$

Subtract each from the number following it:

$$1, \quad 7, \quad 19, \quad 37, \quad 61, \quad 91, \quad 127, \quad \text{etc.}$$

Subtract again:

$$6, \quad 12, \quad 18, \quad 24, \quad 30, \quad 36, \quad \text{etc.}$$

Once more: $6, \quad 6, \quad 6, \quad 6, \quad 6, \quad \text{etc.}$

For cubes we get the row of equal numbers at the third subtraction. With fourth powers we may expect the row of equal numbers to come after the fourth subtraction:

$$0, \quad 1, \quad 16, \quad 81, \quad 256, \quad 625, \quad 1296, \quad 2401, \quad 4096, \quad \text{etc.}$$
$$1, \quad 15, \quad 65, \quad 175, \quad 369, \quad 671, \quad 1105, \quad 1695, \quad \text{etc.}$$
$$14, \quad 50, \quad 110, \quad 194, \quad 302, \quad 434, \quad 590, \quad \text{etc.}$$
$$36, \quad 60, \quad 84, \quad 108, \quad 132, \quad 156, \quad \text{etc.}$$
$$24, \quad 24, \quad 24, \quad 24, \quad 24, \quad \text{etc.}$$

With fifth powers we require five subtractions, and we can even guess what the number will be. For first powers (1, 2, 3, 4, 5, etc.) the constant number is 1.

1st powers	...	...	1
2nd powers	...	...	2
3rd powers	...	...	6
4th powers	...	...	24
5th powers	...	...	?

We can be pretty sure there is a sequence. But what is the sequence? The second number is twice the first, the third is three times the second, the fourth is four times the third;

the fifth is—five times the fourth, that is $24 \times 5 = 120$. Let us check that. The fifth powers are:

$$0, 1, 32, 243, 1024, 3125, 7776, 16,807, 32,768, \text{etc.}$$
$$1, 31, 211, 781, 2101, 4651, 9031, 15,961, \text{etc.}$$
$$30, 180, 570, 1320, 2550, 4380, 6930, \text{etc.}$$
$$150, 390, 750, 1230, 1830, 2550, \text{etc.}$$
$$240, 360, 480, 600, 720, \text{etc.}$$
$$120, 120, 120, 120, \text{etc.}$$

For sixth powers we can now be pretty sure that we shall get constant numbers after the sixth subtraction, and that these will be $120 \times 6 = 720$.

It is not only powers of numbers that can be dealt with like this. Any regular sequence will give us the same kind of result. Thus:

$$1 \times 2, \quad 3 \times 3, \quad 5 \times 4, \quad 7 \times 5, \quad \text{etc.}$$

is a regular sequence; the first number in each product increases by 2 each time and the second number by 1. Multiplied out the sequence is:

	2,	9,	20,	35,	54,	77,	etc.
1st differences:		7,	11,	15,	19,	23,	etc.
2nd differences:			4,	4,	4,	4,	etc.

Here is a sequence with three factors in each product:

$$1 \times 2 \times 3, \quad 2 \times 4 \times 6, \quad 3 \times 6 \times 9, \quad 4 \times 8 \times 12, \quad \text{etc.}$$

The first factor goes up by 1's, the second by 2's, the third by 3's. Multiplied out we have:

	6,	48,	162,	384,	750,	1296,	etc.
1st differences:		42,	114,	222,	366,	546,	etc.
2nd differences:			72,	108,	144,	180,	etc.
3rd differences:				36,	36,	36,	etc.

In writing the fifth powers:

0, 1, 32, 243, 1024, 3125, 7776, 16,807, 32,768, etc.

I noticed that they end in the series 0, 1, 2, 3, 4, etc. and I could not help wondering why. I suspect it has something to do with "five twos are ten". Let us have a look.

Suppose a number ends with the digit n (any number from 0 to 9); and that its fifth power also ends in n. Let us find what the fifth power of $n+1$ ends in:

$$(n+1)^5 = n^5 + 5n^4 + 10n^3 + 10n^2 + 5n + 1$$

(that is a matter of multiplying out)

$$= n^5 + 5n\,(n^3 + 2n^2 + 2n + 1) + 1.$$

Now if n happens to be even $5n$ is an exact number of times 10 (five twos are ten!). If n is odd then the other factor is even (n^3—odd, $2n^2 + 2n$—even, 1—odd; and so the whole factor is even); so again we have five twos are ten; whatever n may be $5n\,(n^3 + 2n^2 + 2n + 1)$ stands for a number which ends in 0.

Look again at $n^5 + 5n\,(n^3 + 2n^2 + 2n + 1) + 1$. We are trying to find the end figure only, so we need not bother with the middle part which is a multiple of 10. We have left $n^5 + 1$. We know that n^5 ends with the digit n; $n^5 + 1$ is one more, and so it must end in $n+1$. That is $(n+1)^5$ ends in $n+1$ (if n^5 ends in n).

See what we have proved: if n^5 ends in n, then $(n+1)^5$ ends in $n+1$. But 0^5 ends in 0; therefore 1^5 ends in 1; therefore 2^5 ends in 2; therefore 3^5 ends in 3; therefore 4^5 ends in 4; and so on.

If we write down the last figures only of the various powers we get:

1st powers:	0, 1, 2, 3, 4, 5, 6, 7, 8, 9, 0, etc.
2nd powers:	0, 1, 4, 9, 6, 5, 6, 9, 4, 1, 0, etc.
3rd powers:	0, 1, 8, 7, 4, 5, 6, 3, 2, 9, 0, etc.
4th powers:	0, 1, 6, 1, 6, 5, 6, 1, 6, 1, 0, etc.
5th powers:	0, 1, 2, 3, 4, 5, 6, 7, 8, 9, 0, etc.
6th powers:	0, 1, 4, 9, 6, 5, 6, 9, 4, 1, 0, etc.
7th powers:	0, 1, 8, 7, 4, 5, 6, 3, 2, 9, 0, etc.
8th powers:	0, 1, 6, 1, 6, 5, 6, 1, 6, 1, 0, etc.
9th powers:	0, 1, 2, 3, 4, 5, 6, 7, 8, 9, 0, etc.
10th powers:	0, 1, 4, 9, 6, 5, 6, 9, 4, 1, 0, etc.

and so the lists go on repeating. This is bound to happen because the fifth line repeats the first line, therefore the 9th, 13th, 17th, 21st, etc. powers give 0, 1, 2, 3, 4, 5, 6, etc.

Suppose we write the sequence of numbers 1 up to, say, 20. We want to find the sum:

$$1+2+3+4+\text{etc.}+19+20.$$

We can do it in a very elegant way, by repeating the sequence in the reverse order:

$$1+ 2+ 3+ 4+\text{etc.} +19+20$$
$$20+19+18+17+\text{etc.} + 2+ 1.$$

We add the terms in pairs, and each pair totals 21. As there are 20 pairs, the total is $21 \times 20 = 420$. But we have added the series twice, so the sum of the series is $420 \div 2 = 210$. We add the first term and the last and multiply by half the number of terms.

When there is an odd number of terms we can get the result quickly by multiplying the middle term by the number of terms:

$$6+7+8= 7 \times 3 = 21,$$
$$8+9+10+11+12 = 10 \times 5 = 50, \quad \text{and so on.}$$

If the number of terms is even we can multiply the sum of the middle pair by half the number of terms:

$$6+7+8+9 = (7+8) \times \tfrac{4}{2} = 30,$$
$$9+10+11+12+13+14 = (11+12) \times \tfrac{6}{2} = 69, \quad \text{and so on.}$$

We are never quite content until we have generalized. Let us write n for any number at all. We want to find:

$$1+2+3+4+\text{etc. up to } +n.$$

Just as before:

$$1+ \quad 2 \quad + \quad 3 \quad + \quad 4 \quad +\text{etc.}+n-2+n-1+n$$
$$n+n-1+n-2+n-3+\text{etc.}+ \quad 3 \quad + \quad 2 \quad +1.$$

Each pair of terms adds up to $n+1$, and as there are n pairs the total is $(n+1) \times n$, or $n(n+1)$. We have added the series twice; the sum we want is therefore:

$$\tfrac{1}{2}n(n+1).$$

Suppose we want the sum of all the numbers up to 100—a tedious task without this help.

$$n = 100, \text{ therefore } \frac{n}{2}(n+1) = \tfrac{100}{2}(100+1)$$
$$= 50 \times 101$$
$$= 5050.$$

The sum of all the numbers up to a million is ($n = 1,000,000$):

$$\frac{1,000,000}{2} \times (1,000,000+1)$$
$$= 500,000 \times 1,000,001$$
$$= 500,000,500,000.$$

The sum of all the numbers up to 1 is ($n = 1$):

$$\tfrac{1}{2} \times (1+1)$$
$$= \tfrac{1}{2} \times 2$$
$$= 1.$$

There is another way in which we can find the sum $1+2+3+4+$ etc. up to $+n$. Suppose we make up our minds that the sum is equal to a certain number of n^3+ a certain number of n^2+ a certain number of $n+$ a number. That is:

$$1+2+3+\text{etc.} +n \equiv An^3+Bn^2+Cn+D,$$

and we have to find what A, B, C and D stand for.

We have to remember that an identity is true no matter what may be the values of the variables.

Write $n=0$. The series is reduced to nothing. So:

$$0=A\times0^3+B\times0^2+C\times0+D$$
$$=D,$$

$D=0$, so there is no number at the end.

Now write $n=1$:

$$1=A\times1^3+B\times1^2+C\times1$$
$$=A+B+C.$$

$n=2$: $\qquad 1+2=A\times2^3+B\times2^2+C\times2,$

or $\qquad\qquad 3=8A+4B+2C.$

$n=3$: $\qquad 1+2+3=A\times3^3+B\times3^2+C\times3,$

or $\qquad\qquad 6=27A+9B+3C.$

We have three equations to find A, B and C:

$$A+\ B+\ \ C=1 \quad (a),$$
$$8A+4B+2C=3 \quad (b),$$
$$27A+9B+3C=6 \quad (c).$$

We can solve these equations by the usual method of algebra:

Multiply (a) by 2: $\quad 2A+2B+2C=2 \quad (d)$.

Take (d) from (b): $\qquad 6A+2B=1.$

Multiply (a) by 3: $\quad 3A+3B+3C=3 \quad (e)$.

Take (e) from (c): $\qquad 24A+6B=3.$

We now have two equations to find A and B:

$$6A + 2B = 1,$$
$$24A + 6B = 3.$$

Multiply the first by 3 and subtract it from the second:

$$6A = 0 \quad \text{or} \quad A = 0.$$

Then $\qquad\qquad 2B = 1 \quad \text{or} \quad B = \tfrac{1}{2}.$

From (a) we now find that $C = \tfrac{1}{2}$. And so the sum is:

$$\tfrac{1}{2}n^2 + \tfrac{1}{2}n = \frac{n}{2}(n+1).$$

We can use the same method to find the sum of the squares: 1^2, 2^2, 3^2, 4^2, etc.

Suppose: $\qquad 1^2 + 2^2 + 3^2 + \text{etc.} + n^2 \equiv An^3 + Bn^2 + Cn + D.$

When $n = 0$: $\qquad\qquad 0 = D.$

When $n = 1$: $\qquad\qquad 1 = A + B + C$

$\qquad\qquad\qquad\qquad$ [since D is 0].

When $n = 2$: $\qquad 1^2 + 2^2 = 8A + 4B + 2C.$

When $n = 3$: $\qquad 1^2 + 2^2 + 3^2 = 27A + 9B + 3C.$

We have three equations to find A, B and C:

$$A + B + C = 1,$$
$$8A + 4B + 2C = 5,$$
$$27A + 9B + 3C = 14.$$

In the same way as before we can find the values of A, B and C:

$$A = \tfrac{1}{3}, \quad B = \tfrac{1}{2}, \quad C = \tfrac{1}{6}.$$

So the sum of the squares is:

$$\tfrac{1}{3}n^3 + \tfrac{1}{2}n^2 + \tfrac{1}{6}n$$
$$= \tfrac{1}{6}n(2n^2 + 3n + 1)$$
$$= \tfrac{1}{6}n(n+1)(2n+1).$$

To test the formula, let us make $n = 4$:

$$1^2 + 2^2 + 3^2 + 4^2 = \tfrac{1}{6} \times 4\,(4+1)\,(8+1),$$

or

$$1 + 4 + 9 + 16 = \tfrac{2}{3} \times 5 \times 9,$$

or

$$30 = 30.$$

We can find the sum of the cubes: $1^3 + 2^3 + 3^3 +$ etc. by putting:

$$1^3 + 2^3 + 3^3 + \text{etc.} + n^3 \equiv An^4 + Bn^3 + Cn^2 + Dn.$$

It should be clear that there is no number at the end, since both sides must equal o when $n = 0$. We have to put $n = 1$, $n = 2$, $n = 3$, $n = 4$; this gives four equations for finding A, B, C and D. Without setting down all the working:

$$A = \tfrac{1}{4}, \quad B = \tfrac{1}{2}, \quad C = \tfrac{1}{4}, \quad D = 0,$$

and so we have:

$$\tfrac{1}{4}n^4 + \tfrac{1}{2}n^3 + \tfrac{1}{4}n^2 = \frac{n^3}{4}\,(n^2 + 2n + 1)$$

$$= \frac{n^2}{4}\,(n+1)^2.$$

We might notice that $\dfrac{n^2}{4}\,(n+1)^2$ is the square of $\dfrac{n}{2}\,(n+1)$. And so:

$$1^3 + 2^3 + 3^3 + \text{etc.} + n^3 = (1 + 2 + 3 + \text{etc.} + n)^2.$$

For the sum of the fourth powers we have to assume:

$$An^5 + Bn^4 + Cn^3 + Dn^2 + En,$$

and we need five equations. The final result is:

$$\frac{n}{30}\,(n+1)\,(2n+1)\,(3n^2 + 3n - 1).$$

Most people have at some time or other worked out fractions of the kind:

$$\cfrac{1}{2+\cfrac{1}{5+\frac{1}{4}}}$$

where you start at the bottom and work upwards, and finally reach $\frac{21}{46}$. The reverse process is much more interesting—changing a fraction like $\frac{21}{46}$ into a continued fraction.

We want the numerator to be 1, so we divide numerator and denominator each by 21:

$$\frac{21}{46}=\frac{\frac{21}{21}}{\frac{46}{21}}=\cfrac{1}{2+\frac{4}{21}}$$

We repeat the process with $\frac{4}{21}$ (dividing numerator and denominator by 4).

$$\cfrac{1}{2+\frac{4}{21}}=\cfrac{1}{2+\cfrac{1}{\frac{21}{4}}}=\cfrac{1}{2+\cfrac{1}{5+\frac{1}{4}}}$$

Look at another example:

$$\frac{11}{37}=\cfrac{1}{3\frac{4}{11}}=\cfrac{1}{3+\cfrac{1}{2\frac{3}{4}}}=\cfrac{1}{3+\cfrac{1}{2+\cfrac{1}{1+\frac{1}{3}}}}$$

Now look at this series of fractions:

$$\frac{1}{3},\ \cfrac{1}{3\frac{1}{2}},\ \cfrac{1}{3+\cfrac{1}{2+1}},\ \cfrac{1}{3+\cfrac{1}{2+\cfrac{1}{1\frac{1}{3}}}}$$

We get the first by ignoring everything after the first 3 in the continued fraction, then everything after the 2, and so on. Let us see how these fractions differ from $\frac{11}{37}$.

$$\frac{11}{37} = \cdot 2973,$$

$\frac{1}{3} = \cdot 3333$ Difference $+ \cdot 0360,$

$$\frac{1}{3\frac{1}{2}} = \frac{2}{7} = \cdot 2857 \qquad\qquad\qquad -\cdot 0116,$$

$$\frac{1}{3\frac{1}{3}} = \frac{3}{10} = \cdot 3 \qquad\qquad\qquad +\cdot 0027,$$

$$\cfrac{1}{3+\cfrac{1}{2+\cfrac{1}{1\frac{1}{3}}}} = \frac{11}{37} = \cdot 2973 \qquad\qquad 0.$$

The differences are alternately above and below $\frac{11}{37}$, and each is nearer to it than the previous one. Hence the fractions are called convergents. We should expect a very good approximation to a fraction when one of the convergents has a small remainder, because we are ignoring a very small quantity when we take that convergent.

Continued fractions have a very small use in enabling us to find approximations to awkward fractions, vulgar or decimal. Take π for example. A very accurate value for π is $3\cdot141592654$. We want to find an approximation to the decimal part.

$$\cdot141592654 = \frac{141592654}{1000000000} \qquad = \cfrac{1}{7+\frac{8851422}{141592654}}$$

$$= \cfrac{1}{7+\cfrac{1}{15+\frac{8821324}{8851422}}} \qquad = \cfrac{1}{7+\cfrac{1}{15+\cfrac{1}{1+\frac{30098}{8821324}}}}$$

$$= \cfrac{1}{7+\cfrac{1}{15+\cfrac{1}{1+\cfrac{1}{293+\frac{2610}{30098}}}}}$$

We could of course go further. By putting 3 in front of each convergent we get a series of approximations to the value of π.

$$\text{(i)} \quad 3\tfrac{1}{7} = \tfrac{22}{7}$$

A quite good approximation, because the fraction we ignore (less than $\tfrac{1}{15}$) is small.

$$\text{(ii)} \quad 3\,\frac{1}{7+\tfrac{1}{15}} = 3\tfrac{15}{106} = \frac{333}{106}$$

Not much better than $\tfrac{22}{7}$; we ignore $\dfrac{1}{1+}$

$$\text{(iii)} \quad 3\,\frac{1}{7+\dfrac{1}{15+1}} = 3\tfrac{16}{113} = \frac{355}{113}$$

A very close approximation; the fraction we ignore is a little less than $\tfrac{1}{293}$.

$$\text{(iv)} \quad 3\,\frac{1}{7+\dfrac{1}{15+\tfrac{1}{293}}} = 3\tfrac{4396}{31065}$$

Very little closer than the last approximation and much more cumbrous.

XIV

The Construction and Solution of Problems

In his book on Lewis Carroll, Walter de la Mare comments on Lewis Carroll's problems. He quotes a friend as saying: "In some cases the construction of the problem is an almost more remarkable performance than its solution, for the author must have foreseen the details as well as the method of their working, otherwise he would not have been able to arrange his figures so that the answers 'come out'."

This statement contains almost every conceivable error that it is possible to make about the construction of problems. In the first place the construction of any ingenious problem is an almost infinitely more remarkable performance than its solution by someone else; solution by the author is part of the process of construction. The author does not usually foresee the details of his problem. In the process of solving it he finds certain awkward points, and he gets over these by introducing suitable numbers. He may find his problem insoluble in its first form and he may have to give further information; one can often detect such added details by the fact that they do not run so smoothly as the rest of the problem. He may find that the problem has two or more answers, and he may add a detail to cut them all out except one. He may disguise some of the information by subtlety of language.

Mr de la Mare quotes one of Lewis Carroll's problems to illustrate his points; and as it equally well illustrates my criticism I give it in full:

"'Five friends agreed to form themselves into a Wine-Company (Limited). They contributed equal amounts of wine, which had been bought at the same price. They then elected one of themselves to act as Treasurer; and another of them undertook to act as Salesman, and to sell the wine at 10 *per cent* over cost price.

'The first day the Salesman drank one bottle, sold some, and handed over the receipts to the Treasurer.

'The second day he drank none, but pocketed the profits on one bottle sold, and handed over the rest of the receipts to the Treasurer.

'That night the Treasurer visited the cellars and counted the remaining wine. "It will fetch just £11," he muttered to himself as he left the cellars.

'The third day the Salesman drank one bottle, pocketed the profits on another and handed over the rest of the receipts to the Treasurer.

'The wine was now all gone; the Company held a meeting, and found to their chagrin that their profits (i.e. the Treasurer's receipts, less the original value of the wine) only cleared 6*d*. a bottle of the whole stock. These profits had accrued in three equal sums on the three days (i.e. the Treasurer's receipts for the day, less the original value of the wine taken out during the day, had come to the same amount every time); but of course only the Salesman knew this.

'(1) How much wine had they bought? (2) At what price?'

In this problem the *number* of friends would seem at first sight to be immaterial; but this is not the case. The answer to the problem as proposed is that 60 bottles were bought at 8*s*. 4*d*. each; but if there had been *four* friends in the Company instead of *five* the answer to the same problem would have been 48 bottles at 10s. each."

Before examining the construction of the problem let us solve it.

Suppose they had x bottles, bought at y shillings each. And suppose the numbers sold on the three days were l, m and n.

There are five unknowns, so we need five equations to solve the problem. We have one equation straight away:

$$l + m + n = x. \qquad (a)$$

The selling-price of a bottle is $\frac{11}{10} y$ (10 per cent or $\frac{1}{10}$ increase on y); the profit on each bottle is to be $\frac{1}{10} y$.

The total profit (after allowing for defalcations) is said to be $\frac{1}{2}x$ shillings (sixpence on each bottle). And so the profit on each day is $\frac{1}{6}x$ (a third of the total profit).

First day: Receipts: $l - 1$ bottles at $\frac{11}{10} y$ each $= \frac{11}{10} y(l - 1)$. Costs: ly.

Profit:
$$\tfrac{11}{10} y(l - 1) - ly = \tfrac{1}{6}x. \qquad (b)$$

Second day: Receipts: $m \frac{11}{10} y - \dfrac{y}{10}$. Costs: my.

Profit:
$$\tfrac{11}{10} my - \dfrac{y}{10} - my = \tfrac{1}{6}x. \qquad (c)$$

Third day: Receipts: $(n - 1) \frac{11}{10} y - \dfrac{y}{10}$. Costs: ny.

Profit:
$$\tfrac{11}{10} y(n - 1) - \dfrac{y}{10} - ny = \tfrac{1}{6}x. \qquad (d)$$

The value of n bottles taken out on the third day is said to be £11 sale price, or £10 ($= 200s.$) cost price.

Hence:
$$ny = 200. \qquad (e)$$

We now have five equations to find the five unknowns. Simplifying them a little we have:

$$l+m+n=x, \qquad (a)$$
$$3ly-33y=5x, \qquad (b)$$
$$3my-3y=5x, \qquad (c)$$
$$3ny-36y=5x, \qquad (d)$$
$$ny=200. \qquad (e)$$

We can get rid of both x and y from (b), (c) and (d) by dividing:

$$\frac{(b)}{(c)}=\frac{3y(l-11)}{3y(m-1)}=\frac{5x}{5x}$$
$$\therefore \quad l-11=m-1 \quad \text{or} \quad m=l-10.$$
$$\frac{(b)}{(d)}=\frac{3y(l-11)}{3y(n-12)}=\frac{5x}{5x}$$
$$\therefore \quad l-11=n-12 \quad \text{or} \quad n=l+1.$$

If we write these values for m and n in the equation:

$$l+m+n=x,$$

we get: $\qquad\qquad 3l-9=x. \qquad (f)$

We can also write the value $n=l+1$ in:

$$ny=200.$$

This gives us: $\qquad (l+1)\,y=200. \qquad (g)$

And we can write 200 for ny in:

$$3ny-36y=5x.$$

This gives us: $\qquad 600-36y=5x. \qquad (h)$

We now have three equations (f, g, h) and three unknowns (l, x, y).

From (f) we have: $\qquad l=\dfrac{x+9}{3}$

Write this value for l in (g):

$$\left(\frac{x+9}{3}+1\right)y=200,$$

or $$(x+12)\,y=600,$$

or $$y=\frac{600}{x+12}$$

Write this value for y in (h):

$$600-\frac{36\times600}{x+12}=5x.$$

Multiplying by $x+12$:

$$600x+7200-21,600=5x^2+60x,$$

or $$5x^2-540x+14,400=0,$$

or $$x^2-108x+2880=0.$$

In factorizing this we have to find two factors of 2880 that add up to 108. Both cannot be odd, therefore both must be even. Hence one factor ends in 0; so that the other must end in 8 (to give a sum of 108). The factors that end in 8 are: 288 (too big), 48 (yes! $48\times60=2880$; $48+60=108$).

Hence $$(x-48)\,(x-60)=0,$$
$$x=48 \text{ or } 60$$
$$y=\frac{600}{x+12}=10 \text{ or } \tfrac{25}{3}$$

Hence the two solutions: 48 bottles at 10s. each and 60 bottles at $\tfrac{25}{3}s$. ($=8s.\ 4d.$) each. Since five people each contributed an equal number of bottles the number of bottles must be exactly divisible by 5. This rules out the first solution, and the correct answer is 60 bottles at 8s. 4d. each.

The solution of the problem is easy enough; it only requires a little care. The interest is in the construction of the problem, and in translating the statements into mathematical terms.

Now this problem almost certainly originated in the whimsical idea of the Salesman drinking bottles of wine and taking the profits on others (not the whole price, because that would be mathematically equivalent to taking the whole bottle). It is equally certain that the number of friends was added as the last touch to the problem—in fact when the necessity for it became apparent, as is shown in the solution. The rate of profit may very well have been put in at the start, with the mental proviso that it might have to be altered if it did not work out well. The three extra unknowns (l, m, n) are neatly hinted at in the problem; their connecting link, $l+m+n=x$, supplies one equation. The arrangement that the profit should be in three equal parts supplies three equations, though a sort of apology is necessary for the fact that only the Salesman knew about this. But we still need a fifth equation. It was for this that the natural but ineffective inspection by the Treasurer was introduced.

I have analysed Lewis Carroll's problem very carefully because it illustrates several points about problems in general, and one kind of problem in particular—a kind of problem depending on oddity of statement rather than on mathematical difficulty or interest. Such problems originate in some point of whimsicality; and it is then only necessary to test whether the problem can be solved or whether some modification is necessary.

Someone thought of the rather odd little idea: 20 silver coins worth 20 shillings, but including no shillings. What are the coins?

Any possible solutions can be found by going through the possibilities systematically. We will investigate the case where only half-crowns, florins and sixpences are included.

In a pound we may have any number of half-crowns from 8 down to 0. 8, 7 and 6 half-crowns are quickly ruled out as possibilities:

half-crowns	florins	sixpences	number of coins
5	3	3	11
5	2	7	14

(As we change each florin to sixpences the number of coins is increased by 3. Thus we have 11, 14, 17, 20 coins. So that there is a solution: 5 half-crowns + 15 sixpences = 20s. and 20 coins).

4	5	0	9

(The number of coins with 4 half-crowns may be 9, 12, 15, 18, 21. No solution.)

3	6	1	10

(Number of coins: 10, 13, 16, 19, 22. No solution.)

2	7	2	11

(Number of coins: 11, 14, 17, 20. Here we have the solution: 2 half-crowns + 4 florins + 14 sixpences = 20s. and 20 coins.)

There are no solutions for 1 half-crown and 0 half-crowns. Thus we have two solutions: 5 half-crowns and 15 sixpences; 2 half-crowns, 4 florins and 14 sixpences. If we state the problem as: "20 coins worth 20 shillings, including half-crowns, florins and sixpences, but no other coins", there is only one solution.

This problem illustrates the advantage of using algebra. Set out in algebraic form we have:

$$x = \text{number of half-crowns},$$
$$y = \text{number of florins},$$
$$z = \text{number of sixpences}.$$

Number of coins $= x + y + z = 20$.

Value of coins $= 2\frac{1}{2}x + 2y + \frac{1}{2}z = 20$ (shillings).

There are two equations and three unknowns. There are an indefinite number of solutions unless we make the proviso that x, y and z are integers. (This is the case with coins.) We can simplify the solution by getting rid of z:

$$x + y + z = 20,$$
$$5x + 4y + z = 40.$$

(Subtracting) $\qquad 4x + 3y = 20.$

Give x the possible values 5, 4, 3, 2, 1, 0 and find the corresponding values of y.

$$x: \quad 5, \quad 4, \quad 3, \quad 2, \quad 1, \quad 0.$$
$$y: \quad 0, \quad 1\tfrac{1}{3}, \quad 2\tfrac{2}{3}, \quad 4, \quad 5\tfrac{1}{3}, \quad 6\tfrac{2}{3}.$$

The two integral values of y are 0 ($x = 5$) and 4 ($x = 2$). These give the solutions:

$$x = 5, \qquad y = 0, \qquad z = 15,$$

and $\qquad x = 2, \qquad y = 4, \qquad z = 14.$

These are the solutions we found before.

Many problems are based on indeterminate equations. The solutions must be limited in some way, usually to positive whole numbers, or there would be an infinite number of solutions. We can make the limitation by dealing with coins, live animals, and other things which usually occur in positive whole numbers.

Take an example at random: 50 animals cost £50. Some cost £5 each, others £1 each, and others 10s. each. How many were there of each kind?

Calling the numbers x, y and z we have:

$$x + y + z = 50,$$
$$5x + y + \tfrac{1}{2}z = 50,$$
or
$$10x + 2y + z = 100.$$

(Get rid of z)
$$9x + y = 50.$$

$$x = 5, \quad 4, \quad 3, \quad 2, \quad 1,$$
$$y = 5, \quad 14, \quad 23, \quad 32, \quad 41.$$

(Add the values of z from $x + y + z = 50$.)

$$z = 40, \quad 32, \quad 24, \quad 16, \quad 8.$$

There are 5 solutions, which is 4 too many. We can alter the prices so as to reduce the number of solutions. Suppose we change the £1 price to £2.

$$5x + 2y + \tfrac{1}{2}z = 50,$$

and the equation for x and y becomes:

$$9x + 3y = 50,$$
or
$$3(3x + y) = 50.$$

There is no solution in integers because 3 is not a factor of 50.

Now make the £1 into £3. The equation for x and y becomes:
$$9x + 5y = 50.$$

$$x = 5, \quad 4, \quad 3, \quad 2, \quad 1,$$
$$y = 1, \quad 2\tfrac{4}{5}, \quad 4\tfrac{3}{5}, \quad 6\tfrac{2}{5}, \quad 8\tfrac{1}{5}.$$

There is only one solution in integers: 5 at £5 + 1 at £3 + 44 at 10s. = £50 and 50 animals.

In such problems we have no guarantee that there is any solution in integers until the equations have been tested. Look at another example:

Peaches can be packed in boxes holding 16 or in baskets holding 15. How many boxes and baskets are needed to hold 100 peaches exactly?

$$16x + 15y = 100.$$

$x=6,$	5,	4,	3,	2,	1,	0,
$15y=4,$	20,	36,	52,	68,	84,	100.

There is no solution in positive integers. But if we change 15 to a number which is a factor of any of the numbers in the second row we get a solution or solutions. If we want only one solution we choose a number that is a factor of only one of the numbers in the second row. Thus 13, 14 and 17 each give one solution only.

To show how a problem may be developed we will consider the following simple example:

A number of counters can be divided exactly into 2's, 3's, 4's, 5's or 6's. How many are there?

The least common multiple of 2, 3, 4, 5 and 6 is:

$$2 \times 2 \times 3 \times 5 = 60.$$

60 is the least possible number; but 120, 180 or any other multiple of 60 satisfies the conditions. We can exclude all solutions except 60 by stipulating the least possible number.

The problem may be made slightly more difficult by adding a remainder of 1 in each case. The answer is of course 61 (or 121, 181, etc.). Or we might have a remainder of 1 with 2, 2 with 3, 3 with 4, etc. The answer is 59 (119, 179, etc.).

We can increase the difficulty by stipulating a remainder of 1 in each case and in addition that the number is to divide

exactly by 7. (Note that 7 is prime to all the other numbers.) We begin by finding numbers which satisfy the first condition, i.e.:
$$61, \ 121, \ 181, \ 241, \ \text{etc.,}$$
and then we find which of these is exactly divisible by 7. The smallest possible number is the next in the series—301. To find other numbers which satisfy the conditions we should have to add multiples of 60×7 (= 420). 420 divides exactly by all the numbers concerned and therefore makes no difference to the remainders. The sequence of solutions is: 301, $301 + 420 = 721$, $301 + 2 \times 420 = 1141$, etc.

There is a sudden increase in difficulty when all the remainders are made different, and there may be no solution at all. There is no number, for example, which gives a remainder 2 when divided by 3 and a remainder 1 or 3 or 4 when divided by 6. ($3x + 2$ gives a remainder 2 when divided by 3. When $3x + 2$ is divided by 6 the remainder is 2 when x is even, and 5 when x is odd.) We can make sure of finding a solution by making the divisors prime to each other. Here is an example at random:

What number gives a remainder 4 when divided by 9, a remainder 7 when divided by 13, and a remainder 12 when divided by 17?

(i) The number must be a multiple of 9 with 4 added. It must be in the series:
$$9+4, \quad 18+4, \quad 27+4, \quad 36+4, \quad \text{etc.}$$

(ii) It must be a multiple of 13 with 7 added. Suppose we deduct 7 from each of the terms of the series in (i), the remainder must be a multiple of 13:

$$9+4=13, \ 18+4=22, \ 27+4=31, \ 36+4=40, \ \text{etc.}$$
$$(-7) \qquad 6, \qquad\quad 15, \qquad\quad 24, \qquad\quad 33, \ \text{etc.}$$

(The series of course increases by 9's.)

Which term divides exactly by 13? The remainders, after division by 13, are: 6, 2, 11, 7, 3, 12, 8, 4, 0. So we want the ninth term, that is $9 \times 9 + 4 = 85$. $9 \times 13 = 117$ is exactly divisible by 9 and 13, so the addition of any multiple of 117 does not affect the remainders. Thus we get a series of numbers which satisfy the first two conditions. This series is:

$$85, \quad 85 + 117 = 202, \quad 85 + 2 \times 117 = 319, \text{ etc.}$$

(iii) Proceed in the same way with the series:

$$85, \quad 202, \quad 319, \quad 436, \text{ etc.}$$

Deduct the third remainder (12):

$$73, \quad 190, \quad 307, \quad 424, \text{ etc.}$$

($117 \div 17 = 6 + R\ 15$. Hence keep adding 15 to the remainder and deduct 17 when possible. $73 \div 17 = 4 + R\ 5$. $5 + 15 - 17 = 3$; $3 + 15 - 17 = 1$; $1 + 15 = 16$; and so on.) The remainders are: 5, 3, 1, 16, 14, 12, 10, 8, 6, 4, 2, 0. There are 12 before we reach 0, so we want the twelfth number in the series:

$$85, \quad 85 + 117, \quad 85 + 2 \times 117, \quad \text{etc.},$$

that is $85 + 11 \times 117 = 1372$. 1372 is the lowest number which satisfies all three conditions. $9 \times 13 \times 17 = 1989$ is exactly divisible by 9, 13 and 17, and so makes no difference to the remainders. Thus any term of the following series satisfies the conditions:

$$1372, \quad 1372 + 1989 = 3361, \quad 3361 + 1989 = 5350, \text{ etc.}$$

Thus:
$$5350 \div 9 = 594 + R\ 4,$$
$$5350 \div 13 = 411 + R\ 7,$$
$$5350 \div 17 = 314 + R\ 12.$$

Here is another example of the elaboration of a comparatively simple problem into a much more difficult one.

How many squares of all sizes are there in this drawing?

Of squares like $ABCD$ there are obviously 4^2.

The next-sized square is $AEFG$. We can move this on to $BHIJ$ and $EKLF$—3 squares. We can also move it down to MN and to OP; thus giving 3^2 squares of this size. (For each move down there are 3 moves across.)

Similarly we find the number of squares like $AHQM$—2^2. And finally the large square—1^2.

The total number of squares is: $1^2+2^2+3^2+4^2$. If there had been 5 squares in each side the total would be: $1^2+2^2+3^2+4^2+5^2$. And so on.

We can increase the difficulty by asking for the number of oblongs (including squares) of all kinds:

(i) Oblongs 1 square deep and 1 square long—4×4; 1 deep and 2 long (4 down and 3 across)—4×3; 1 deep and 3 long—4×2; 1 deep and 4 long—4×1.

Total 1 square deep: $4 (4+3+2+1)$.

(ii) 2 deep and 1 long—3×4; 2 deep and 2 long—3×3; 2 deep and 3 long—3×2; 2 deep and 4 long—3×1.

Total 2 squares deep: $3 (4+3+2+1)$.

(iii) 3 deep and 1 long—2×4; 3 deep and 2 long—2×3; 3 deep and 3 long—2×2; 3 deep and 4 long—2×1.

Total 3 squares deep: $2 (4+3+2+1)$.

(iv) 4 deep and 1 long—1×4; 4 deep and 2 long—1×3; 4 deep and 3 long—1×2; 4 deep and 4 long—1×1.

Total 4 squares deep: $4+3+2+1$.

Complete total: $(4+3+2+1)^2$
$$= 1^3+2^3+3^3+4^3.$$

The next stage is to consider a cube made up of small cubes. The number of cubes of all sizes is readily found by the same kind of reasoning as for the square, except that we can now move in three directions. The total is:

$$1^3+2^3+3^3+4^3.$$

The final stage of the problem is to find the number of oblong prisms of all shapes and sizes. These may be 1, 2, 3 or 4 of the small cubes in height. We have already found the number, 1 cube high, in the lowest layer:

$$(4+3+2+1)^2.$$

Since there are four layers there are altogether:

$$4 (4+3+2+1)^2$$

of these.

Consider prisms 2 cubes high. In the lowest layer there are $(4+3+2+1)^2$, as before. These can be raised one place or two places, so that the total is $3 (4+3+2+1)^2$.

It will now readily be seen that the total of prisms of all shapes and sizes is:

$$(4+3+2+1) (4+3+2+1)^2 = (4+3+2+1)^3.$$

We can generalize the results by dividing each side of the square or cube into n equal parts.

Number of squares: $1^2 + 2^2 + 3^2 + \text{etc.} + n^2$.

Number of oblongs: $1^3 + 2^3 + 3^3 + \text{etc.} + n^3$,

or $\qquad\qquad (1 + 2 + 3 + \text{etc.} + n)^2$

$$= \frac{n^2}{4}(n+1)^2.$$

Number of cubes: $1^3 + 2^3 + 3^3 + \text{etc.} + n^3$.

Number of prisms: $(1 + 2 + 3 + \text{etc.} + n)^3$

$$= (\text{if you like}) \ \frac{n^3}{8}(n+1)^3.$$

We will now devise one or two problems to illustrate various methods. We will begin with a "bottles of wine" problem.

Three merchants agree to sell bottles of wine at a profit of 25 per cent. A pays the cost of one-third of the bottles, B pays the profit on two-thirds of the bottles, and C pays —?

Let us imagine 120 bottles bought at 12s. per bottle and sold at 15s. per bottle. A's payment is $40 \times 12s. = 480s.$ B's payment is $80 \times 3s. = 240s.$ C's payment brings the total to $120 \times 12s. = 1440s.$; that is, C pays $720s. = \pounds 36$.

We now have one equation. Call the number of bottles x; let them be bought at y shillings each, and sold at z shillings each. We have:

A's payment $+ B$'s payment $+ C$'s payment $=$ total cost,

$$\frac{xy}{3} + \tfrac{2}{3}x\,(z - y) + 720 = xy.$$

This works down to: $2xy - xz = 1080$.

We want two more equations. One that suggests itself is

that the total sale price (1800) is $7\frac{1}{2}$ times B's payment (240). This gives the equation:

$$xz = \tfrac{15}{2} \times \tfrac{2}{3}\,(xz - xy)$$
$$= 5xz - 5xy,$$

or $\qquad 4xz - 5xy = 0.$

We have now got two equations:

$$2xy - xz = 1080,$$

and $\qquad 4xz - 5xy = 0.$

We can solve them to get xy and xz. We find:

$$xy = 1440; \quad xz = 1800.$$

$$\frac{xy}{xz} = \frac{y}{z} = \frac{1440}{1800} = \text{(among other things)} \ \frac{12}{15}$$

If we can limit y and z in some way, so that only the values 12 and 15 can be used, we shall have completed the problem. We have $\frac{12}{15} = \frac{24}{30} = \frac{4}{5}$. If we make the profit a whole number of shillings, less than 6 and more than 1, the thing is done, without a third equation. Let us see how it works out.

Three merchants buy a stock of wine and agree to sell it at an exact number of shillings profit per bottle. A suggests 1s. profit per bottle. "Too little", says C. B suggests 6s. per bottle. "Too much", says C. Finally all agree to C's proposal. "I will pay for a third of the bottles", says A. "I will pay enough to make up the profit on two-thirds of the bottles", says B. "I will pay the rest", says C, and he puts down £36. When the wine was all sold B observed that the total receipts were exactly $7\frac{1}{2}$ times what he had paid. How many bottles were there?

Notice that the remarks about the profits come at the beginning, not because this point was foreseen, but because any intriguing oddity gives a good start to the problem. If it came at the end it would spoil the problem.

I have just thought of this:

Suppose we give A 30 counters, B 45, C 60. We also give E 40, F 55 and D 70. I have chosen these numbers because I want E, F and D each to have 10 more than one of the others—but we are not going to say which.

So far we might state the problem: B has 15 more counters than A, and C has 15 more than B. D, E and F each have 10 counters more than one of the first three, a different one in each case. How many counters has each?

Of course there is not yet sufficient information to solve the problem. We notice that D has as many as A and E together. How far does that carry us?

A	B	C
x	$x+15$	$x+30$

D, E, F each has $x+10$, $x+25$ or $x+40$. D has the sum of A's share and E's, that is either $2x+10$, $2x+25$ or $2x+40$. The last case is impossible because no one has more than $+40$. Hence D has either $2x+10$ or $2x+25$.

We also know that D has either $x+25$ or $x+40$. (He cannot have $x+10$ because he has more than E.)

We have now two ways of expressing what D has:

$$2x+10 \quad \text{or} \quad 2x+25,$$
and
$$x+25 \quad \text{or} \quad x+40.$$

So we can say:

$$2x+10=x+25 \quad \text{or} \quad 2x+10=x+40,$$
$$\therefore \quad x=15 \qquad\qquad \therefore \quad x=30,$$

or else
$$2x+25=x+25 \quad \text{or} \quad 2x+25=x+40,$$
$$\therefore \quad x=0 \qquad\qquad \therefore \quad x=15.$$

A can therefore have 0, 15 or 30 counters. We want one solution only, and we can get this by making *A's* number even. Very well:

A small estate is divided among six people. *A* gets an even number of pounds; *B* gets £15 more than *A*; *C* gets £15 more than *B*. *D*, *E* and *F* each get £10 more than one of the first three, a different one in each case. *D* gets as much as *A* and *E* together. What was the value of the estate?

Some problems are simply arithmetical facts stated in language that hides the facts. For example:

$$17 \times 19 = 323; \quad 323 \text{ farthings} = 6s. \ 8\tfrac{3}{4}d.$$

So, if we buy a number of articles for 6s. 8$\tfrac{3}{4}$d., and exclude the price of $\tfrac{1}{4}$d. each, we must have bought 17 at 19 farthings each ($=4\tfrac{3}{4}d.$) or 19 at 4$\tfrac{1}{4}$d. each. It is not difficult to devise wording that will exclude all possibilities but one. Thus:

I bought a number of toys for 6s. 8$\tfrac{3}{4}$d. I happened to notice that if I had paid $\tfrac{1}{4}$d. more for each toy I could still have bought an exact number of them for 6s. 8$\tfrac{3}{4}$d. How many toys did I buy? (The only solution is 17.)

Here is a kind of problem that depends on a mathematical fact:

$$x^3 - x = x\ (x^2 - 1) = (x - 1)\ x\ (x + 1).$$

So that the difference between a number and its cube is the product of three consecutive numbers.

$$24 \times 25 \times 26 = 15,600.$$

So we can say: The difference between a number and its cube is 15,600. What is the number?

In solving the problem we have to find the three consecutive numbers whose product is 15,600.

$$15,600 = 2^4 \times 3 \times 5^2 \times 13.$$

The cube root of 15,600 is between 20 and 30, so we look for numbers between 20 and 30. 13 is a factor of one of the numbers, so that one must be 26. This quickly gives 25 and 24 as the other factors. 25 is the number.

$$25^3 - 25 = 15,625 - 25 = 15,600.$$

We have already met the relation:

$$1^3 + 2^3 + 3^3 + 4^3 = (1 + 2 + 3 + 4)^2$$

and so on for any number of the cubes of consecutive numbers. Let us see if we can do anything with it.

$$1^3 + 2^3 + \text{etc.} + 12^3 = (1 + 2 + 3 + \text{etc.} + 12)^2$$
$$1^3 + 2^3 + \text{etc.} + 9^3 = (1 + 2 + 3 + \text{etc.} + 9)^2$$

Subtracting:

$$10^3 + 11^3 + 12^3 = (1 + 2 + \text{etc.} + 12)^2 - (1 + 2 + \text{etc.} + 9)^2$$
$$= 78^2 - 45^2$$
$$= (78 + 45)(78 - 45)$$
$$= 123 \times 33.$$

$123 - 33 = 90 = 2 \times 45$. (That is obvious, because 123 is 45 more than 78, and 33 is 45 less than 78.)

Now: Piles of dice can be built up exactly into three cubes, the number of dice in the edges being consecutive numbers. They can also be spread out to form an exact oblong with 90 more dice in the length than in the width. How many dice are there?

To show that the problem is soluble let us take another number. Suppose there are 42 more dice in the length than in the width (but we cannot take a number at random).

Half of 42 is 21, and

$$21 = 1 + 2 + 3 + 4 + 5 + 6.$$

Hence the three cubes are 7^3, 8^3, 9^3.

$$7^3 + 8^3 + 9^3 = 1584$$
$$= (x-21)(x+21)$$
$$= x^2 - 21^2.$$
$$\therefore \quad x^2 = 1584 + 21^2 = 2025,$$
$$x = 45.$$

The sides of the oblong are: $45 + 21 = 66$, and $45 - 21 = 24$.
Finally:
$$66 \times 24 = 1584.$$

There is a well-known problem which illustrates the construction of a problem by elaboration, and also the method of solving such a problem by starting at one end or the other. The problem is:

Three niggers robbed an orchard, and decided to share the apples in the morning. In the night one nigger got up and divided the apples into three equal parts. There was one apple over; he threw it away, took his part, and went to sleep. Later on a second nigger did the same thing. Again there was one apple over; he threw it away, took his part, and went to sleep. The third nigger also did the same thing, threw away the apple that was over, took his part, and went to sleep. In the morning no one mentioned the incidents of the night. They divided the remaining apples into three equal parts, threw away one which was over, and each took a third. How many apples were there?

The possibility of a solution depends of course on the fact that we must have a whole number of apples at each division. We can start at either end.

(i) Suppose each gets x apples at the last division. There were $3x + 1$ apples before this division. This was the $\frac{2}{3}$ left by the third nigger. Before his division there were:

$$\tfrac{3}{2}(3x+1) + 1.$$

Before the second division there were:

$$\tfrac{3}{2}\{\tfrac{3}{2}\,(3x+1)+1\}+1.$$

And before the first division there were:

$$\tfrac{3}{2}\,[\tfrac{3}{2}\,\{\tfrac{3}{2}\,(3x+1)+1\}+1]+1$$
$$=\frac{81x+65}{8}$$
$$=10\tfrac{1}{8}x+8\tfrac{1}{8}.$$

This must be a whole number, and x also is a whole number. Hence $\tfrac{1}{8}x+\tfrac{1}{8}$ is a whole number. (We can discard $10x+8$.) Hence $\tfrac{1}{8}x$ can be $\tfrac{7}{8}$ (to make up $\tfrac{1}{8}$ to 1), $\tfrac{15}{8}$ (to make up $\tfrac{1}{8}$ to 2), etc. So that:

$$x=7,\ 15,\ 23,\ 31,\ \text{etc.}$$

When $x=7$, $10\tfrac{1}{8}x+8\tfrac{1}{8}=79$, the lowest possible solution. Other solutions are: $79+81\ (=160)$, $79+2\times81\ (=241)$, etc. Note that $81=3^4$. So that the addition of any multiple of 81 makes no difference to any of the remainders.

(ii) Suppose there were x apples at the start. (This begins at the other end.) The first nigger took $\tfrac{1}{3}\,(x-1)$ and left $\tfrac{2}{3}\,(x-1)$.

The second left: $\tfrac{2}{3}\{\tfrac{2}{3}\,(x-1)-1\}$.

The third left: $\tfrac{2}{3}\,[\tfrac{2}{3}\{\tfrac{2}{3}\,(x-1)-1\}-1]$.

In the final division each got:

$$\tfrac{1}{3}\,(\tfrac{2}{3}\,[\tfrac{2}{3}\{\tfrac{2}{3}\,(x-1)-1\}-1]-1)$$
$$=\frac{8x-65}{81}.$$

This must be a whole number.

Hence $\qquad 8x-65=\text{multiple of }81,$

$$8x=\text{multiple of }81+65$$
$$x=\text{multiple of }10\tfrac{1}{8}+8\tfrac{1}{8}.$$

This is the point we arrived at before, except that x is now the original number of apples. It is easy to see that the multiple must be 7, $7+8$, $7+2 \times 8$, etc. A general solution is:

$$x = (8n+7) \ 10\tfrac{1}{8} + 8\tfrac{1}{8}.$$

When $n=0$, we have $x=79$. When $n=1$, $x=160$. And so on, with the same series of solutions as before.

Suppose there were 4 niggers, we should have the same sort of solution with $\tfrac{4}{3}$ instead of $\tfrac{3}{2}$, and an additional division:

$$\tfrac{4}{3} \left[\tfrac{4}{3} \left\{ \tfrac{4}{3} \left(\tfrac{4}{3} \overline{4x+1} + 1 \right) + 1 \right\} + 1 \right] + 1$$
$$= \tfrac{1024}{81} x + \tfrac{781}{81}$$
$$= 12\tfrac{52}{81} x + 9\tfrac{52}{81}$$

Hence $\dfrac{52x}{81} + \dfrac{52}{81}$ must be a whole number. I.e. $52 \left(\dfrac{x}{81} + \dfrac{1}{81} \right)$ must be a whole number.

$$\therefore \quad x = 80 \quad \text{or} \quad 80 + 81n.$$

When $x = 80$: $\quad 12\tfrac{52}{81} x + 9\tfrac{52}{81} = 1021.$

This is the least possible solution. Other solutions are $1021 + 1024$, $1021 + 2 \times 1024$, etc. $1024 = 4^5$, so that the addition of multiples of 1024 makes no difference to the remainders on division by 4, 4^2, 4^3, 4^4 or 4^5.

For 5 niggers we should have:

$$\frac{5^6}{4^5} x + \frac{5^5 + 5^4 \times 4 + 5^3 \times 4^2 + 5^2 \times 4^3 + 5 \times 4^4 + 4^5}{4^5}$$

must be a whole number. And so on. I.e. $15\tfrac{265}{1024} x + 11\tfrac{265}{1024}$ must be a whole number.

$$\therefore \quad x = 1023 \quad \text{or} \quad 1023 + 1024n.$$

The number is $15{,}621$, or $15{,}621 + 15{,}625$, etc.

Now if we examine the solutions we shall see that the minimum solutions are:

For 3: $79 =   81 - 2 = 3^4 - 2.$
For 4: $1{,}021 = 1{,}024 - 3 = 4^5 - 3.$
For 5: $15{,}621 = 15{,}625 - 4 = 5^6 - 4.$

We can add:

For 6: $6^7 - 5.$
For 7: $7^8 - 6.$
For 8: $8^9 - 7.$ And so on.
For n: $n^{n+1} - (n-1).$

I had no intention of doing that when I started to work out the problem, but one is led on. These solutions are minimum solutions. The general solution is:

$$n^{n+1} - (n-1) + mn^{n+1},$$

where m and n are any positive integers.

XV

Scales of Notation

WE are so used to our decimal notation in writing numbers that we almost regard it as a sort of natural product, rather than as the mathematical device that it is. The illusion is heightened by the fact that the decimal scale is in universal use amongst civilized peoples.

If we write a number at random, say 64,238, and consider what this means:

$$6 \text{ stands for } 6 \times 10^4,$$
$$4 \text{ stands for } 4 \times 10^3,$$
$$2 \text{ stands for } 2 \times 10^2,$$
$$3 \text{ stands for } 3 \times 10^1,$$
$$8 \text{ stands for } 8 \times 10^0,$$

and if we add ·927:

$$9 \text{ stands for } \tfrac{9}{10} \text{ or } 9 \times 10^{-1}$$
$$2 \text{ stands for } \tfrac{2}{100} \text{ or } 2 \times 10^{-2}$$
$$7 \text{ stands for } \tfrac{7}{1000} \text{ or } 7 \times 10^{-3}$$

It is the sequence of powers that is the genius of the system; the 10 is a mere accident of the system. The 10 probably arose from the facts that we have 10 fingers and that we use the fingers in counting. If we had happened to have 8 fingers we might have had an octonary scale, running in powers of 8. Or if we had had 12 fingers we might have had a duodecimal scale, in powers of 12.

As soon as we try to deal with scales other than the scale

of 10, we meet the difficulty that nearly all the number names are based on the scale of 10. Twenty is two tens; a hundred is 10 squared; a thousand is 10 cubed; and so on. But two eights is 16 (ten and six); there is no 2-eight name corresponding to twenty—the 2-ten name. Eight squared is 64 (6 tens and 4); there is no simple name for 8 squared corresponding to hundred for 10 squared.

The difficulty is one of naming only. The ease with which we work in the decimal system is not due to any mathematical superiority of that system, but partly to the fact that our language has been framed to suit that system and partly to the fact that we learn the multiplication tables in the decimal system.

Suppose we choose another scale, say the scale of 5. In this scale 2341 would stand for:

$$2 \times 5^3 + 3 \times 5^2 + 4 \times 5 + 1.$$

(Just as in the scale of 10 it stands for:

$$2 \times 10^3 + 3 \times 10^2 + 4 \times 10 + 1.)$$

In the scale of 5, 10 stands for 5; there is no need for any digit higher than 4. Just as there is no need for any digit higher than 9 in the scale of 10; or any digit higher than 11 in the scale of 12.

There are actually traces of the scale of 12 (the duodecimal scale) in our language. We have the dozen (= 12), the gross (= 12²), and the great gross (= 12³). But we are short of two digits to use this scale properly; we need digits for ten and eleven. We can get over this difficulty by writing *t* for ten and *e* for eleven. In the duodecimal scale, 9*tet* stands for:

9 great gross, ten gross, eleven dozen, and ten.

Let us turn one of the multiplication tables into the scale of 12:

$$7 \times \ 1 = 7$$
$$7 \times \ 2 = 12 \ (1 \text{ dozen and } 2)$$
$$7 \times \ 3 = 19 \ (1 \text{ dozen and } 9)$$
$$7 \times \ 4 = 24 \ (2 \text{ dozen and } 4)$$
$$7 \times \ 5 = 2e \ (2 \text{ dozen and eleven})$$
$$7 \times \ 6 = 36 \ (3 \text{ dozen and } 6)$$
$$7 \times \ 7 = 41 \ (4 \text{ dozen and } 1)$$
$$7 \times \ 8 = 48 \ (4 \text{ dozen and } 8)$$
$$7 \times \ 9 = 53 \ (5 \text{ dozen and } 3)$$
$$7 \times \ t = 5t \ (5 \text{ dozen and ten})$$
$$7 \times \ e = 65 \ (6 \text{ dozen and } 5)$$
$$7 \times 10 = 70 \ (7 \text{ dozen}).$$

That is the table we should have to learn if we used the scale of 12. Apart from the language difficulty it is no more difficult than the corresponding table, the one we have to learn, for the scale of 10.

The table of 6's in the duodecimal scale is very much like the table of 5's in the decimal scale:

$$6 \times 1 = 6$$
$$6 \times 2 = 10 \ (1 \text{ dozen})$$
$$6 \times 3 = 16 \ (1 \text{ dozen and } 6)$$
$$6 \times 4 = 20 \ (2 \text{ dozen})$$
$$6 \times 5 = 26 \ (2 \text{ dozen and } 6), \quad \text{and so on.}$$

The table of elevens in scale 12 is very much like the table of nines in scale 10. (Eleven is one less than twelve, and nine is one less than ten.)

$$e \times 1 = e$$
$$e \times 2 = 1t \ (1 \text{ dozen and ten})$$
$$e \times 3 = 29 \ (2 \text{ dozen and } 9)$$
$$e \times 4 = 38 \ (3 \text{ dozen and } 8)$$
$$e \times 5 = 47 \ (4 \text{ dozen and } 7)$$
$$e \times 6 = 56, \quad \text{and so on.}$$

The duodecimal table of 3's runs: 3, 6, 9, 10, 13, 16, 19, 20, 23, 26, 29, 30.

The duodecimal table of 9's runs: 9, 16, 23, 30, 39, 46, 53, 60, 69, 76, 83, 90.

Our money system includes no fewer than four different scales of notation—the 4 scale which we use in changing farthings to pence, the duodecimal scale for pence and shillings, a 20 scale for shillings and pounds, and the decimal scale for pounds. So far from being a disadvantage this mixture of scales is far superior to a purely decimal scale for most purposes.

We have a binary (2) scale for small weights: $\frac{1}{4}$ oz., $\frac{1}{2}$ oz., 1 oz., 2 oz., $\frac{1}{4}$ lb., $\frac{1}{2}$ lb., 1 lb. Such a scale is of course ideal for weighing. When we come to stones we again use a binary scale—$\frac{1}{4}$ stone, $\frac{1}{2}$ stone, stone. The method of weighing for tons and hundredweights renders a binary scale unnecessary. The measure of capacity is almost entirely binary: $\frac{1}{4}$ pt., $\frac{1}{2}$ pt., 1 pt., 1 quart, 1 gallon (=4 quarts), 1 peck, 1 bushel (=4 pecks).

We are used in England to a great variety of scales of notation, and all the world is used to the scales employed in reckoning time. It comes as something of a shock to hear that a centesimal division of the right angle, long ago proposed in France, has recently been decreed in Germany. The right angle is to be divided into a hundred degrees each of a hundred minutes; it must annoy pedants that four right angles to the circle must inevitably persist—surely it ought to be ten, if not a hundred.

I wonder who has thought out the implications of the change in Germany? The change will certainly provide a little employment in recalculating and reprinting the trigonometrical tables—sines, cosines, tangents, etc. There would be further employment in making new instruments marked with the new degrees—protractors, sextants, theodolites, and so on. Angles come into so many calculations—surveying, astronomical, nautical—that the price to be paid for a bit of pedantry seems rather high. No one has yet explained what advantage division into hundredths has over division into ninetieths. It is not as if we ever had to multiply or divide angles—we don't; or even as if we had to add or subtract many angles; the cases that arise are extremely simple. Having measured an angle we seldom use the measurement directly; we use one of the trigonometrical ratios—sine, cosine, tangent, etc. We are therefore ultimately dependent on the radian, and one of the new degrees is ·015708 radian, which has no obvious superiority over the equivalent in radians of one of our degrees—·01745 radian. Will some decimalist, or should I say centesimalist, explain exactly what the advantage is? Will the Minister for Education in Germany explain why he has made the decree? We sit in arithmetical gloom, waiting for enlightenment.

The angles of an equilateral triangle will be 66⅔ of the new degrees instead of the 60° we are all familiar with. Indeed, as regards divisibility, 100 is a poor substitute for 90. 90 has the factors:

2, 3, 5, 6, 9, 10, 15, 18, 30, 45,

whereas 100 has:

2, 4, 5, 10, 20, 25, 50.

We throw away the important factor 3, for the dubious advantage of an extra 2. We throw away also the parallel with clock measurements (minutes and seconds) which is useful in astronomical work. And I can perceive no gain whatsoever.

I have tried to make it clear that the decimal scale of notation has no arithmetical superiority over other scales of notation; in the matter of divisibility it is far inferior to the duodecimal scale. The decimal scale of notation holds its place as a vested interest, and not for any arithmetical reason. Language has been framed to suit the decimal system; there are vast numbers of arithmetical calculations in the decimal scale; its use is common throughout the civilized world. For these reasons it is highly improbable that for ordinary calculations the inferior decimal system will ever be abandoned. But that is no reason why we should permit the encroachment of the inferior system upon our admirable system of money, weights and measures. Uniformity may cost more than it is worth.

CAMBRIDGE: PRINTED BY W. LEWIS, M.A., AT THE UNIVERSITY PRESS

Printed in the United States
By Bookmasters